Trust, New Technologies and Geopolitics in an Uncertain World

I0127942

Pascaline Gaborit

Trust, New Technologies and Geopolitics in an Uncertain World

PETER LANG

Bruxelles · Berlin · Chennai · Lausanne · New York · Oxford

Bibliographic Information published by the Deutsche Nationalbibliothek
The Deutsche Nationalbibliothek lists this publication in the Deutsche Nationalbibliografie; detailed bibliographic data is available online at http://dnb.d-nb.de.

Library of Congress Cataloging-in-Publication Data
A CIP catalog record for this book has been applied for at the Library of Congress.

Cover image: Pascaline Gaborit (assisted by CANVA/AI)
Cover Design by Peter Lang

The book was written as part of projects funded by the European Union. CRIC: Climate Resilient and Inclusive Cities - and AI4DEBUNK. This project has received funding from the European Union Horizon Innovation Actions under grant agreement N°101135757. The views and opinions expressed in this article are those of the authors alone and do not necessarily reflect those of the project partners, the European Union or the European Commission. Neither the project partners nor the European Commission can be held responsible for these views and opinions.

ISBN 978-3-0343-5845-3 (Print)
ISBN 978-3-0343-5846-0 (E-PDF)
ISBN 978-3-0343-5847-7 (E-PUB)
DOI 10.3726/b22919
D/2025/5678/20

© 2025 Peter Lang Group AG, Lausanne (Switzerland)
Published by P.I.E. PETER LANG s.a., Bruxelles (Belgium)

info@peterlang.com

Table of Contents

Figures

Contributions and Acknowledgments

This book was written with the help of two contributors or research assistants, **V. Rao and Joen Martinsen**, whom I thank for their help and intelligent contributions.

I thank the project teams and the Pilot4dev board members for their valuable inputs, my partner and my children for their support.

The book was written thanks to the cross-fertilization of research and analysis from various projects, some of which were funded by the European Union: CRIC: Climate Resilient and Inclusive Cities and AI4DEBUNK Grant Agreement No. 101135757.

The views and opinions expressed in this article are solely those of the author and do not necessarily reflect the positions of the project partners, the European Union, or the European Commission. The project partners and European Commission bear no responsibility for these views and opinions. This book, being a synthesis of various analyses and research, is not a project deliverable.

Artificial Intelligence was used to verify information, search for contextual elements, help with reformulation, sentence structuring, and translation.

Abstract: In an era of unprecedented global uncertainty, trust has become a critical factor shaping the relationships between nations, institutions, and individuals. This book, *Trust, New Technologies and Geopolitics in an Uncertain World*, offers a timely and in-depth exploration of how trust is being tested and transformed in the face of rapidly shifting geopolitical landscapes. From the fragility of democratic systems to the challenges posed by new technologies, disinformation, and climate change, this book delves into the most pressing issues of our time. By examining the intersections of trust with key areas such as the Sustainable Development Goals (SDGs) and ongoing global conflicts, this work provides valuable insights for policymakers, scholars, and anyone seeking to understand the complexities of today's world. Whether you are concerned with cybersecurity, the impact of hybrid threats, or the role of trust in international diplomacy, this book offers a comprehensive yet accessible framework to navigate these challenges. It sheds light on how rebuilding and reimagining trust could be key to addressing the geopolitical uncertainties that define our age.

Author: Pascaline Gaborit is a researcher and consultant for international projects. She is director of the NGO/Think Tank Pilot4dev. She is the author of numerous articles and books on subjects related to governance, resilience, climate adaptation, conflict, and new technologies. She is also passionate about marine life.

Contributors: V. Rao and J. Martinsen

Introduction

Imagine a world without trust—a world of fear and insecurity, stripped of rules and certainty.

Such a world would be unbearable to live in. Society would collapse, and mere survival would become a constant struggle. Conversely, imagine a world where trust in the system is absolute and unshakable—such a world would resemble a totalitarian regime, oppressive and soul-crushing for the individual. When people become overly trusting and unaware of risks, they become easy targets for manipulation and vulnerable to threats—in other words, gullible. That is why trust is not merely an ideal to be achieved—it is a fundamental element, a factor and a parameter that helps us analyze complex issues. We will try to explore in this book how trust can deepen our understanding of current geopolitical situations.

Trust is a powerful, multifaceted force. It shapes not just our emotions and perceptions but also the credibility of people, institutions, and even currencies. As the world grows more interconnected, trust has become a cornerstone of cooperation, essential for navigating our increasingly complex global landscape.

The rising emphasis on trust today can be traced to the growing uncertainty and the legitimacy crises faced by institutions worldwide. As traditional, magical, and religious explanations lose ground and education spreads, our understanding of risk becomes more nuanced, elevating the importance of trust. Where once tradition fostered passive confidence, this reliance has evolved into a more active, decisive form of trust in our modern world.

In an age where science and technology dictate much of our lives, trust is more crucial than ever. We rely on everything from transportation and safety devices to technological innovations. With the rise of digitization and unprecedented technological advancements, trust becomes even more layered, amplifying the complexities of our daily existence.

Trust is at the heart of every meaningful interaction, whether between individuals or larger collectives. It's a process that requires embracing vulnerability—whether with oneself, others, or even institutions—especially in times of uncertainty.

Beyond personal relationships, trust is woven into the fabric of all human endeavors. From knowledge-sharing to the management of political, scientific, and economic power, trust influences our ability to work together, make decisions, and drive progress.

Confidence is often thought of as the expectation of a nearly certain outcome. But its opposite, doubt, highlights the uncertainty we face in every aspect of life. Trust, in its truest sense, is about embracing this uncertainty—finding a strategy to move forward despite the unknown. It serves as a simplifying force, helping us navigate social complexities, especially in the face of ignorance or uncertainty. Above all, trust is the glue that holds social and contractual order together, enabling cooperation and interaction in an unpredictable world.

The end of the 1990s marked the collapse of the communist bloc, the opening up to democracy and the spread of the market economy, raising new hopes for civil societies worldwide (Fukuyama, 2005). In the 2000ies and 2010ies, pro-democracy movements flourished in many parts of the world: in Georgia, in Ukraine, with the Arab Springs (2010–2012) and the Jasmin revolutions, even in Hong Kong (2014) with the "occupy central" movement. Movements for peace and democracy also spread to Iran and Belarus. Few of these pacific revolutions were able to lead to a shift toward sustainable democratic systems.

The **Rose Revolution in Georgia** took place in November 2003. It was a peaceful protest movement that led to the resignation of President Eduard Shevardnadze and the rise to power of Mikheil Saakashvili. The revolution was triggered by widespread allegations of electoral fraud in the parliamentary elections of November 2, 2003. Mass demonstrations, led by opposition leaders and civil society groups, culminated in Shevardnadze's resignation on November 23, 2003. The movement was characterized by its nonviolent nature and the symbolic use of roses, representing a peaceful transition of power. It marked a significant shift toward democratic reforms and closer ties with the West.

The Orange Revolution in Ukraine in 2004 is a notable exception as it succeeded to change the elections' results and to change the regime. The Orange Revolution in Ukraine took place between November 2004 and January 2005 following the fraudulent presidential election of 2004. It was a series of protests and political events triggered by widespread allegations of election fraud in favor of Viktor Yanukovych. In response, mass demonstrations led to a revote in December 2004, which resulted in the victory of Viktor Yushchenko, the opposition candidate. The change of regime and new elections were only confirmed after the "Occupy Maidan" or "EuroMaidan' pro Europe protests in 2013 and 2014 and the final defeat of Yanukovych.

The **Arab Springs** took place between December 2010 and mid-2012, though their effects continued for years. They began with the Tunisian Revolution in December 2010, following the self-immolation of Mohamed Bouazizi in protest against government corruption and police brutality. The movement quickly spread across the Middle East and North Africa (MENA), leading to uprisings, protests, and in some cases, civil wars in countries like Egypt, Libya, Syria, Yemen, and Bahrain. While some regimes were overthrown, others responded with repression.

The Gezi movement in Turkey, beginning in June 2013, marked a pivotal moment in Turkey's contemporary history. What started as opposition to an urban redevelopment project threatening Gezi Park in Istanbul rapidly evolved into a widespread popular uprising against civil liberty restrictions. Though ultimately suppressed, the movement heralded a new era of activism for fundamental freedoms, especially among youth and middle-class citizens. Turkey has since witnessed further waves of protest, including demonstrations following Istanbul Mayor Ekrem İmamoğlu's imprisonment on March 19, 2025.

The Occupy Central movement in Hong Kong, also known as the Umbrella Movement, took place in 2014. It began on September 28, 2014, and lasted until December 15, 2014. The protests were triggered by Beijing's decision to pre-screen candidates for the 2017 Chief Executive election, which activists saw as a violation of democratic principles. Protesters occupied key areas of the city, demanding universal suffrage and greater political freedom.

With the **Green Movement of 2009** and the *Woman, Life, Freedom* **protests of 2022,** Iranian activists, intellectuals, and ordinary citizens have repeatedly challenged authoritarian rule, demanding political freedoms, human rights, and democratic governance.

Did the protesters overestimate their ability to challenge oppressive regimes, placing too much trust in other citizens and in their own strength? Or have new technologies fundamentally reshaped the dynamics of power, altering the balance between resistance and repression?

More recently, the emergence of unforeseen tensions, the onset of Russia's war of aggression against Ukraine in February 2022, and the ongoing conflicts in the Middle East since 2023—marked by human rights violations—have raised new questions. The rising uncertainty is amplified with the rise of technologies, able to undermine democratic principles and the rule of law. Western democracies simultaneously experience a question of trust and trustworthiness when their basis fundamentals are questioned by polarized narratives and the rise of populists' movements. The 2024 U.S. elections and Donald Trump's nomination have sent shockwaves across the globe, raising unprecedented questions.

Are we now witnessing a major step backward for democracies and for the world's stability? Is the world increasingly fragmented? Have we entered a world of uncertainties where the guardians of international law have lost authority? Will international relations come down to a balance of power between the great powers? And finally, what will be the future of civil societies in a chess game where they are being increasingly ignored when not manipulated?

According to experts, the world "has become grounded in geopolitical risks." Geopolitical tensions are escalating with the rivalry between blocks and with the war of aggression on Ukraine. Few sustainable solutions seem to emerge from this escalation of risks. Energy and climate continue to be polarizing issues, with global progress lacking in energy transition (S&P 2024). Climate change is increasingly worrying as the coordinated efforts to respect the Paris Agreement, as well as the global movement toward SDGs seem to be disabled by increasing tensions and distrust (Chapter 4). The situation has become even more concerning since the U.S. decision to withdraw from the Paris Agreement. Globalization as a way to bring peace and prosperity is increasingly questioned since the COVID-19 pandemic. "At times, 2024 has felt like a pivotal moment as the balance of world power appears to tilt on its axis: Conflict, trade tensions, and economic turbulence are contributing to a more uncertain geopolitical global framework" (WEF 2024).

History has been shaped by battles, revolts, revolutions, and developments. The twentieth century was marked by two world wars, the opposition of two blocs during the Cold War, and the race toward nuclear and thermonuclear weapons. These events defined an era of intense conflict and rapid techno-logical advancements aimed at gaining military and ideological superiority. However, although apparently more peaceful the twenty-first century is characterized by uncertainty, the rise of multipolar powers, and the spread of new technologies capable of controlling citizens and minds. Never in history has it been possible to control populations so effectively and com-prehensively. Instruments such as the internet, created to enhance freedom and facilitate communication, have been transformed into tools of power by both governments and malicious actors. In recent years, hybrid attacks have become more frequent and severe, evolving into powerful geopolitical tools alongside other emerging technologies (**Chapter 2**). At the same time, the relatively stable and cooperative global order of the past two decades is giving way to a more turbulent and fragmented landscape (WEF, 2024).

Another major challenge is the rise of **disinformation**, which threatens democratic institutions and global stability (**Chapter 3**). Meanwhile, efforts to address **climate change** and achieve the **SDGs** are facing significant setbacks, raising concerns about long-term sustainability (**Chapter 4**).

In this increasingly uncertain world, **trust** emerges as a critical factor in navigating crises and rebuilding cooperation. A deeper analysis of trust dynamics can provide valuable insights into potential pathways for mitigating conflicts and fostering stability (**Chapter 5**).

But faced with a more fragmented world, experts recognize that distrust is one of the starting points of the current situation, which is a context of increasing tensions worldwide: "The starting point must be to recognize that distrust is, in the short and medium term at least, a baked-in feature of geopolitical reality [...] To lessen the risks presented by a fracturing world, the international community must manage the distrust so that it does not prevent collaboration or escalate to conflict" (WEF 2024). But how do we dig into the complexity of trust and distrust in the current geopolitical situation? How do we use and parameter the analysis on trust to understand the current trends and situations in international relations?

Trust and Uncertainty or Change/Transformation

As we have seen earlier, there is an increasing perception that "the cooperative international order is being replaced by a more turbulent and fragmented global landscape." Digitization and new technologies are transforming the world. Does it mean that trust will be disabled or transformed into a kaleidoscopic phenomenon? Will trust between actors turn into a schizophrenic reality when we look at all the invisible wars linked with trade and cyber threats?

For Barber, trust is represented by the expectations that social actors have of each other. These expectations involve beliefs that social transactions allow the continuation and fulfillment of a moral and social order considered natural (Barber, 1983). The rule of law is one of the conditions of democracy: it organizes political competition (by making it more secure) and reduces attempts at arbitrary power (Sa'adah, 2006:305)·

Society is the result of its decisions, which themselves refer to discussions, conflicts and transactions through which, in an always provisional and unstable manner, changes are pursued that move in the direction of greater diversification, growing flexibility, a loosening of norms, symbolic systems and social constraints (Touraine, 1973).

Western European societies, as studied by sociologists Emile Durkheim, Georg Simmel, Maurice Halbwachs, and Norbert Elias,[1] share many commonalities due to their rapid social changes. These societies have undergone major transformations over the past seven decades, leading to the erosion of once-fundamental social categories—social classes, nuclear families, and gender roles.. Before the 1960s, people generally lived within secure social structures, clearly distinguishing friends from foes, with serious consequences for those who broke societal rules. This pattern persisted through subsequent decades, even as social movements emerged in the 1970s.

In modern societies and with the emergence of the digital era, everyone is faced with strangers, online communities managed by algorithms and anonymous crowds. This makes it harder to calculate the expectations of others.

In societies marked by risk and uncertainty, individuals and groups increasingly delegate trust to experts and specialists. However, as social categories evolve—particularly the distinction between "us" and "others"—and as relationships become more fluid, the need for social trust becomes even more critical. Yet, this trust is frequently undermined, as reflected in the rise of nationalism, populism, and polarizing movements across Europe, which are fueled by distrust and contestation.

---

[1] At the end of the nineteenth century and in the twentieth century.

Indeed, it is possible to confirm that the effects of economic, legal and social transformations concretely modify and sometimes overturn relations between groups and individuals undermine hierarchies and undermine authorities (Lagroye, 1997). The digitization of societies has brought a new revolution, as we will detail in Chapter 3. These changes do not go unnoticed by those who undergo them and come in parallel with economic transformations, social and special inequalities, social exclusion and economic inflation.

Any transformation has different paces and different effects, and we are experiencing rapid changes, in particular with the rise of Artificial Intelligence (AI) and climate challenges, as we will discuss in chapters 3 and 4. Any transition clock has three different rhythms: the lawyer's hour is the shortest; legal changes can be implemented in a few months. The economist's hour is longer. But the longest hour is that of the citizen, who must transform internalized habits, mental attitudes, cultural codes, value systems, and discourses. This can take several years (Dahrendorf, 1990: 3).

In her speech commemorating the 50th anniversary of the Treaty of Rome on March 25, 2007, Angela Merkel made the following remarks about the European Union:: *"Building Trust takes decades, but it can be undermined overnight."*[2] "Trust is challenging to build yet easily broken (Sa Adah 2006: 321)." Does fostering trust in public institutions require a long-term effort, while its erosion happens more swiftly? More fundamentally, can trust truly be destroyed, or is it an intrinsic human need that persists but evolves through shifts in perception?

Trust in institutions is indeed a critical issue that Chapter 1 will address.

The International Community: Balancing Trust and Expectations

In today's fragmented world, or "world disorder," the issues of SDGs and climate change revive questions about the role of the international community. This community remains symbolized by the United Nations, despite recent

---

[2] Statement quoted in *"Le Monde"* newspaper, March 29, 2007, p. 2. She thus reiterated the importance of joint efforts by EU Member States to strengthen European cohesion and values. <https://acrobat.adobe.com/id/urn:aaid:sc:EU:f2279027-e1cd-44b1-81de-0b0efdcc7fd9>

criticism about how global divisions affect the Security Council. Trust in the international community extends beyond the local level, creating a delicate balance between independence and interference, between expectations and results, between populations and actors.

The role of UN actors has been described in these words:

> Wherever he goes, he is like a prisoner of the expectations that these begging populations have of the United Nations and of this exalted fiction: the international community. These expectations legitimize his organization, they are its raison d'être. *And yet, in one way or another, it tries in vain to reduce these expectations, to contain this inevitable disappointment, and to force people to rediscover their own possibilities.*[3]

Kant, in proposing a forum for the representation of states, was visionary in advocating dialogue for peace—a concept still relevant today within international organizations such as the UN (Kant, 1795). However, in practice, like any organization, the UN faces power struggles and conflicting interests among its members and other stakeholders. The extension of the Security Council permanent members to countries like India and the place of the European Union are still debated.[4] The organization often struggles to assert its decisions effectively, which typically emerge from compromise. Trust depends heavily on people's expectations and creating expectations—whether among the general population or key players—can ultimately undermine that trust.

The recent closure of the U.S. Agency for International Development (USAID) and the reduction of the budget from the United States will inevitably shake international organizations. In a sweeping effort to slash the U.S. budget deficit, the Trump administration, along with the unofficial

---

[3] Ignatieff M. 2001.
[4] The United Nations Security Council (UNSC) was established in 1945 following World War II as one of the six principal organs of the United Nations (UN), tasked with maintaining international peace and security. It initially had 11 members, but its composition was expanded to 15 members in 1965. The Council consists of five permanent members (P5)—China, France, Russia (formerly the Soviet Union), the United Kingdom, and the United States—who hold veto power, and ten non-permanent members elected for two-year terms. Over the decades, the UNSC has played a crucial role in conflict resolution, peacekeeping operations, and imposing sanctions, though it has often faced criticism for power imbalances, inaction due to veto use, and its lack of reform.

Department of Government Expenditure (DOGE) led by Elon Musk, has indeed moved to dismantle USAID—placing nearly all employees on leave and freezing foreign aid programs. Security personnel have blocked agency staff from accessing their offices, plunging billions in foreign aid into uncertainty and halting critical humanitarian efforts. Court decisions are still pending in this, but the damage may already be done.

Earlier the COP29 discussions in Baku in November 2024, which focused on funding a climate solidarity mechanism, were not celebrated as a success, as they were exacerbating tensions and eroding trust in the climate negotiations process.[5] "Climate negotiations saboteurs," whether connected to the fossil fuel industry or not, have reduced the topic to a mere financial transaction—a magical thinking that suggests money alone could ensure prosperous crops, halt climate change, and spontaneously lead to the construction of infrastructure and hospitals.

Conversely, the instrumentalization of the international community by political actors in particular can strengthen specific actors, as when actors use the latter's actions to divert them to their own advantage, or to castigate its failings. Finally, let's be honest, even the economy depends on trust.

Trust and the Economy

"Confidence cannot be bought in the marketplace"

— (Arrow, 1974).

The notion of trust first emerged in economics. In the early 1970s, economists began examining trust's role in economic development. They observed that

---

[5] The 2024 United Nations Climate Change Conference (COP29) in Baku, Azerbaijan, concluded with an agreement to provide $300 billion annually by 2035 to assist developing nations in addressing climate change impacts. This figure fell significantly short of the $1.3 trillion per year that developing countries had advocated for, leading to widespread dissatisfaction among these nations. The negotiations were further strained by geopolitical tensions, including the recent U.S. presidential election, which cast doubt on future American climate commitments. Additionally, the host nation's heavy reliance on fossil fuels and its controversial human rights record undermined trust in the process. Collectively, these factors contributed to the perception that COP29 failed to deliver the ambitious climate action and financial support necessary to effectively combat global warming.

rational calculations alone could not fully explain cooperation. Consider a game theory scenario where Alex and Ben exchange items (a watch for a chair): Ben could maximize profit by keeping both items, refusing to give the chair after receiving the watch. This dynamic makes cheating seem more advantageous than cooperation, as Alex bears too great a risk of loss. Though signing a contract can reduce Alex's risk, even if a judge enforces the contract, this process cannot be fully explained by purely rational calculations of economic behavior (Laufer, 2000).

How, then, can we explain exchange, cooperation, and trade? This is where reputation enters the picture. Ben has an incentive to cooperate to maintain a good reputation and continue future exchanges with Alex (Kreps, 1990). However, reputation is not merely an economic concept—it is fundamentally social.

Trust in economic terms was equally theorized by Arrow in 1974, who described it as a necessary lubricant for social relations (Arrow, 1974). In studies of industrial and innovative districts during the 1990s, trust was viewed as a commodity that enables and sustains solidarity among players (Hanssen, 1992). This metaphor of trust as cement or glue captures its binding nature. While economic cooperative behavior remains central to understanding trust, the relationship between trust and cooperation presents a chicken-and-egg dilemma: does the need for trust drive cooperation, or does the need for cooperation create trust?

The idea of economic confidence has now entered everyday language. It is intuitively accepted that economic life requires trust.

The economic system involves numerous players and mechanisms—including monetary value, bank rates, and growth—where information is not always transparent to key stakeholders. In such circumstances, trust and confidence in the system's long-term benefits help overcome this information gap. The 2008 subprime crisis demonstrated how trust and distrust can significantly impact financial markets and influence the global economy. Over the past decade, particularly since the COVID-19 pandemic, trust in globalization has eroded, leading to trade conflicts and a resurgence of "national preference." This shift profoundly impacted international relations. The situation became particularly evident during the global shortage of face masks—and subsequently lack of vaccines—when it became clear that even wealth and influence could not guarantee reliable supply chains. This raised a critical question: should Western nations and developing countries continue to risk such heavy dependence on China and continue to trust the current globalization system?

Trust and International Relations

Trust and international relations have become an increasingly studied topic (Kydd, 2005, Haukkala et al., 2018, Willet et al., 2023). However, few authors have explored beyond the realm of interpersonal trust between world leaders (Brewer, 2017). Analysis of trust between leaders was particularly prominent during the Cold War, enabling observers to detect subtle signs of tension or appeasement in leaders' public statements and attitudes. Far from obsolete, this approach to analyzing trust relationships could illuminate the dynamics between Presidents Donald Trump and Vladimir Putin, or even between Elon Musk and Donald Trump following the formation of the U.S. government in January 2025.

Example on How Interpersonal Trust is Impacting Geopolitics

The best-known example of the impact of interpersonal trust on international relations is the relationship between **U.S. President Ronald Reagan and Soviet leader Mikhail Gorbachev during the Cold War.** Their rapport and mutual trust played a crucial role in easing tensions between the two superpowers. This trust facilitated negotiations that led to important arms control agreements, such as the 1987 Intermediate-Range Nuclear Forces Treaty (INF).

A stark example of deteriorating interpersonal trust occurred during the **meeting in the Oval Office in Washington, D.C., between Ukrainian President Volodymyr Zelensky and U.S. President Donald Trump on Friday, February 28, 2025.** Zelensky faced open humiliation and accusations of ingratitude and inappropriate attire. The meeting, which ended with a canceled joint press conference, came to be regarded as a dark day for Ukraine and the free world. The tense exchange saw Trump and Vice President J. D. Vance openly criticize Zelensky for allegedly exploiting U.S. support—an incident that revealed the fragility of both the transatlantic alliance and support for Ukraine.

In the Russian Federation, other more recent examples can be seen in a climate of heightened mistrust. The relationship between **Vladimir Putin and Yevgeny Prigozhin**, in particular, was based on mutual interests but marred by deep mistrust, ultimately leading to Prigozhin's mysterious death in August 2023.

As the head of the Wagner Group, a mercenary force serving Russia's geopolitical ambitions, Prigozhin was once considered a loyal enforcer of Kremlin interests—from Ukraine to Syria and Africa. However, tensions escalated as Prigozhin's public defiance and ambition began to challenge Putin's authority. The June 2023 Wagner mutiny, when Prigozhin led his forces in an armed rebellion against Russia's military leadership, exposed a crack in Putin's power structure. Although the revolt was swiftly halted through backchannel negotiations, Prigozhin's betrayal shattered any remaining trust between the two. While he was momentarily allowed to retreat, few believed he was truly safe. Just two months later, Prigozhin's private jet exploded mid-flight, a fate widely suspected to have been orchestrated by the Kremlin as a warning to other potential challengers. This episode reaffirmed the ruthlessness of Putin's grip on power, where even long-time allies are not immune to retribution when they step out of line. The Putin-Prigozhin saga serves as a stark reminder of the brutal and paranoid nature of power struggles within authoritarian regimes. This shows that interpersonal trust plays a role in geopolitics.

Several authors highlight the distinction between psychological and rational trust (Michel, 2012, Michel, 2016, Haukkala et al., 2018). Psychological trust stems from attachment and historical ties, while rational trust is rooted in strategic and energy interests. Realpolitik advocates favor rational trust, emphasizing quantifiable and specific objectives, whereas proponents of psychological trust focus on diplomacy, image, influence, culture, and soft power.

Geopolitics is the study of how geography, politics, economics, and power dynamics intersect and influence global relations and decision-making. It examines how physical and human geography—such as borders, natural resources, strategic locations, and demographics—shape the behavior and strategies of states and non-state actors on the global stage. Geopolitics considers competition for influence, control, and security among nations, often influenced by historical legacies, cultural ties, and economic interdependencies. In an increasingly interconnected world, the scope of geopolitics extends beyond territorial disputes to encompass issues like climate change, technological dominance, cyber warfare, and the management of global commons, making it a vital framework for understanding international conflicts and cooperation in an era of uncertainty. Geopolitics provides a broad framework for studying trust

Trust theory has been applied to three different schools of international relations: (1). Realpolitik views trust primarily as a security concern, (2) liberal theory frames it as a function of trade, and (3) constructivist theory sees it as part of a network of perceptions (Wheeler 2011).

Trust is paradoxical because it is both pragmatically based (enabling cooperation) and, in this sense, involves calculation or anticipation, but it also implies faith that exceeds all justification (Giddens, 1991, Möllering, 2001). Trust can be analyzed both from the perspective of rational choice theory (prisoner theory, liberal economic authors) and from a cognitive or subjective point of view.

Another key hypothesis explored in this book is that international tensions arise from pivotal moments of trust-building and trust-breaking. The two primary perspectives on trust—the rational dimension and the subjective or cognitive dimension—offer distinct lenses through which to understand conflicts, shedding light on different underlying mechanisms. Today, we are witnessing a rapid and profound erosion of trust at the international level, a phenomenon unmatched since the end of the Cold War.

Another paradox of trust in a risky situation is to increase the risk level: it is paradoxical in a situation of risk (e.g., armed attack) that trusting (i.e., acting as if the risk were lower or non-existent) actually adds an additional risk to the first risk: that of trusting. This situation can be exemplified by the current position of European NATO members. By placing their trust in the alliance, particularly in Article 5, which establishes a solidarity mechanism for collective defense in the event of an attack—they risk becoming overly reliant on this framework. This reliance may lead to a diminished focus on strengthening their own capacities, potentially leaving them vulnerable in scenarios of attacks where the alliance's support is delayed or impossible. By trusting, we can substitute a new risk for the old one (Sztompka 1999).

Excessive trust can eradicate doubts and fears—and in this sense, as Anthony Giddens puts it, remains *"blind trust"*—whereas conflict resolution and democratic systems typically involve a different type of trust: conditional trust, which grants the benefit of the doubt but can be withdrawn when necessary (Lebow). According to Lebow, even rationalist authors in international relations acknowledge that deterrence rests on the expectation that adversaries will refrain from using nuclear weapons and that deterrence can be enforced (Lebow 2013). "Realists describe the international system as anarchical without any embedded norms. They nevertheless recognize that trust lies at the core of strategies to deter and compel certain behaviors when dealing with allies and foe alike. Target states must believe that you will carry out your threats if they fail to act as your demand" (Lebow, 2013). This applies to today's fragmented geopolitical order, where "hot spots" represent areas of tension with risks of nuclear escalation (Ukraine, Middle East, Korea). This situation contradicts the liberal theory of an emerging international order marked by exchange and trade in a relatively peaceful and stable environment. As we have seen, international relations cycle through periods of trust-building and erosion. Recent events suggest we are currently in a phase of deteriorating trust.

It is also important to consider what Wheeler calls the "security dilemma" in international relations, particularly in the context of the potential use of nuclear, chemical, thermodynamic, or biological weapons. Wheeler describes this dilemma as the paradox in which states, in an effort to enhance their own security, take actions (such as military build-ups) that may be misinterpreted by others, thereby increasing the risk of escalation. He emphasizes the need to "perceive the motives behind the military intentions of others and to be responsive to the potential complexity of those intentions [...] it also refers to the ability to understand the role that fear may play in their attitude and behavior, including, crucially, the role that one's own action may play in provoking that fear" (Wheeler, 2011). Donald Trump's **"shock and awe"** strategy could be seen in this light. However, this approach appears to completely disregard the reactions of other countries, as evidenced by the confusion and lack of understanding expressed by several nations in response to his impulsive statements.

This security dilemma historically enabled a stable equilibrium and explained the balance of terror during the Cold War through mutual nuclear deterrence. However, in today's era of hybrid warfare, the balance of power has likely shifted. There is now less transparency regarding the military capabilities of major powers, partly due to rapid technological advances, making it harder to understand other nations' intentions and capabilities.

This erosion of the security dilemma and balance of terror could trigger escalations with unpredictable consequences.

The type of trust depends on the type of relationship, but trust is necessary for any cooperation between individuals and between groups (Luhmann, 1979, Seligman, 1997, Hardin, 2004). As we have seen in the beginning of our introduction, without trust, there can be no living together, no cooperation. The world would be hell. But trust is also based on subjective feelings and interpretations. It is based on beliefs, attitudes, feelings and expectations (Vuorelma, 2018, Haukkala, 2018, Forsberg, 2018). In cases of extreme mistrust, every gesture, statement and action is interpreted as insincere or manipulative. This distrust affects any possibility of dialogue, cooperation and joint action.

While choosing a single definition is challenging, we propose our own patchwork definition of trust for this book:

Trust is a multifaceted concept: it is (1) a feeling that enables exchange and cooperation, (2) a mechanism for reducing social complexity and coping with uncertainty, and (3) a mental process that allows us to accept/ignore risks and dangers.

Objective and Structure of the Book

Our intention in this book is to go beyond the study of trust as a concept, but to apply it to the current trends in international relations: fragmentation, uncertainties, new technologies, tensions, and difficulties of democracy.

It will address the following points:

– The question of Trust in Democracy (Chapter 1)
– The challenges of new technologies, connectivity and artificial intelligence (Chapter 2)
– The Challenges and Threats of Disinformation (Chapter 3)
– The question of the Dialogue on Climate Change and Sustainable Development Goals (Chapter 4)
– Trust as a parameter in the current Crisis (Conclusion)

The chapters can be read sequentially or independently, depending on the reader's interest.

We have selected these sectors for two main reasons: (1) they are critical in current geopolitical relations, and (2) they align with our scope of study.

Our methodology combines documentary research, key stakeholder interviews, conference participation, and focus group discussions. Through this approach, we contribute to research projects like the CRIC Climate Resilient and Inclusive Cities project and the AI4DEBUNK project on disinformation. These two projects are funded by the European Union.

The research is also based on my work and analysis as a consultant for international organizations, and as the Director of Pilot4dev.[6] The research on trust and conflicts also originated from my Ph.D. research on "Trust in post conflict societies" which was published in 2009 in French (Gaborit 2009).

Limitations of the Research

One of the key limitations of this research lies in the **rapidly evolving nature of geopolitical dynamics**, which poses significant challenges to fully capturing the complexities of trust in global affairs. Geopolitical landscapes, influenced by shifting alliances, emerging threats, and technological advancements, are constantly in flux. This fluidity makes it difficult to offer definitive conclusions, as the issues explored today may quickly be altered by new developments. Additionally, the scope of this research is constrained by the inherent **difficulty in accessing reliable information** on sensitive security issues, such as **cyberattacks** and **hybrid threats**. Governments and institutions often limit the availability of crucial data due to national security concerns, leading to gaps in understanding how these threats shape international trust and cooperation.

Furthermore, the **time constraints** imposed on this work have necessitated a focused examination of only a select number of critical topics. As a result, the analysis centers primarily on key chapters: **democracy, new technologies, disinformation, SDGs** and **climate**, and **trade conflicts**. While these areas are highly relevant to current geopolitical concerns, other equally important topics may remain underexplored such as the impact of Terror Attacks and of Crime and Criminal Activities on Trust. The rapidly changing global

---

[6] <www.pilot4dev.com>

context, combined with limited time and access to sensitive information, therefore narrows the breadth of the research, though the chosen focus areas provide a strong foundation for understanding the intersection of trust and geopolitics in today's uncertain world.

Conclusion and Interest of the Book

Today's world economy faces many challenges. Trade is slowing due to inflation and new policies aimed at protecting sovereignty. Technological innovation is booming, bringing challenges created by connectivity and AI (See Chapter 2). Climate change requires governments and businesses to collaborate to reduce GHG emissions. The SDGs are losing momentum (see Chapter 4). Democracies are weakened and struggling to maintain their principles (Chapter 1), especially in a global context of disinformation (Chapter 3).

Some consider trust to be a cement (Hanssen, 1992: 71) or an invisible institution (Dupuy et al., 1997). However, whether the need for trust leads to cooperation, or whether the need for cooperation fosters trust, remains an enigma for me and for other authors. Our final chapter (conclusion) will explore possible trust as a parameter to understand and address current challenges.

In an era of unprecedented global uncertainty, trust has become a critical factor shaping the relationships between nations, institutions, and individuals. This book, *Trust, New Technologies and Geopolitics in an Uncertain World*, offers a timely and in-depth exploration of how trust is being tested and transformed in the face of rapidly shifting geopolitical landscapes. From the fragility of democratic systems to the challenges posed by new technologies, disinformation, and climate change, this book delves into the most pressing issues of our time. By examining the intersections of trust with key areas such as the SDGs and ongoing global conflicts, this work provides valuable insights for policymakers, scholars, and anyone seeking to understand the complexities of today's world. Whether you are concerned with cybersecurity, the impact of hybrid threats, or the role of trust in international diplomacy, this book offers a comprehensive yet accessible framework to navigate these challenges. It sheds light on how trust construction or destruction could explain the geopolitical uncertainties that define our turbulent time.

References

Arrow K. *The limits of organization*, New York, W. W. Norton & Company, 1974.

Barber B. *The Logic and Limits of Trust*, New Brunswick, Rutgers University Press, 1983.

Bernoux P. and Servet J. M. *La construction sociale de la confiance*, Paris, éditions Montchrestien, 1997.

Bianco W. T. *Trust, Representatives and constituents*, Michigan Press of University, 1994.

Bourdieu P. *La misère du monde*, Paris, Le Seuil, 1993.

Braithwaite V. and Levi Margaret (Dir), *"Trust and governance,"* New York, Russel Sage foundation, trust collection, no. 1, 1998. 386 p.

Brewer, P. R., Gross K., and Vercellotti T., "Trust in International Actors," in Eric M. Uslaner (ed.), *The Oxford Handbook of Social and Political Trust*, Oxford Handbooks, 2018; online edn, Oxford Academic, Jan 10, 2017, <https://doi.org/10.1093/oxfordhb/9780190274801.013.32>

Cappella J. N. and Jamieson K. H. *Spiral of Cynism*, New York, University Oxford Press, 1997.

Dahrendorf R. 1990, quoted in Tonkiss F.; Passey A. (Dir), *Trust and civil society*, London, Macmillan Press, 2000.

Damien R., Lazzeri Ch. (Dir). *Confiance et conflit*, Besançon, Presses universitaires de France Comté, 2006, 394p.

Deutsch M., "Trust and suspicion," *Journal of conflict resolution*, 1958, vol. 2, n° 4, pp. 265–279.

Dunn J. *Trust and political agency*, Diego Gambetta, 1988.

Earle T. and Cvetkovich G. T. *Social Trust: Toward a Cosmopolitan society*, New York, Praeger editions, 1995.

Finc A., Nayrou F. and Pragier G. *La haine: haine de soi, haine de l'autre, haine dans la culture*, Paris, PUF, 2005.

Forsberg T. "Taking Strock of the trust study in International Relations," in Haukkala H., van de Wetering C., Vuorelma J. *Trust in International Relations, Rationalist, Constructivist, and Psychological Approaches*, Routledge, 2018.

Fox J. "Towards a dynamic theory of ethno-religious conflict," *Nations and nationalisms* n°5, ASEN, 1999, pp. 431–463.

Fukuyama F. *Trust: The social virtues of the creation of prosperity*, New York, Free Press Paperbacks, 1995, 457 p.

Gaborit P. *Restaurer la confiance après un conflit civil*, L'Harmattan, 2009 a.

Gaborit P. "La confiance après un conflit ou la confiance désenchantée," in Bertho A., Gaumont-Prat H. et Serry H. Colloque international *La confiance et le conflit*, Université Paris Vincennes Saint Denis, 2009 b.

Gaborit P. *Learning from Resilience Strategies in Tanzania, an Outlook of International Development Challenges*, Peter Lang, 2021.

Gaborit P. "Resilience and Climate Disaster Management in Cities: Transformative Change and Conflicts," *Journal of Peacebuilding & Development*, special issue, Nov. 2022, <https://doi.org/10.1177/15423166221128793>, 2022 (a).

Gaborit P. "Climate Adaptation to Multi-Hazard Climate-Related Risks in Ten Indonesian Cities: Ambitions and Challenges," *Climate Disaster Risk*, 2022, Vol. 37, 100453 <https://doi.org/10.1016/j.crm.2022.100453> 2022 (b).

Gaborit P. (Ed). *Climate Adaptation and Resilience: Challenges and Potential solutions. Anticipatory governance, Planning and Dialogue*, Peter Lang, 2022 c.

Gambetta D. "Can we trust trust?," in D. Gambetta (Dir). *Trust: Making and Breaking Cooperative relations*, Oxford, Basil Blackwell, 1988.

Gellner, 1995 quoted by Tonkiss F.; Passey A. Collective work, *Trust and civil society*, London, Macmillan Press, 2000, p. 2.

Giddens A. *Modernity and self-identity*, Stanford, Stanford University Press, 1991.

Girard R., *La violence et le sacré*, Paris, Éditions Bernard Grasset, Coll. Pluriel, 1972.

Govier T. and Verwoerd W. "Trust and the problem of National Reconciliation", *Philosophy of the Social sciences*, Vol. 32, n°2, June 2002, pp. 178–205.

Gurr T. *Minorities at risk*, Washington, Peace of Institute, 1993.

Hansen N. M. "Competition, trust and reciprocity in the development of Innovative regional milieu," *Regional Science*, 1992, 71.

Hardin R. *Trustworthiness.*, New York, Russel Sage Foundation, 1996.

Hart E. J. *Democracy and Distrust: a theory of Judicial Review*, Editions Harvard University Press, 1988.

Hassner P. *La violence et la paix*, Paris édition du Seuil, 2000.

Haukkala H., Saari S. "The cycle of mistrust in EU-Russia Relations," in Haukkala H., van de Wetering C., Vuorelma J., Routledge, 2018.

Haukkala H., van de Wetering C., Vuorelma J. *Trust in International Relations, Rationalist, Constructivist, and Psychological Approaches*, Routledge, 2018.

Hufschmidt B. and Damien R. (Dir). *Bachelard: confiance raisonnée et défiance rationnelle*, Besançon, Presses universitaires de Franche Comté, 2006.

Ignatieff M. *Human Rights as Politics and Idolatry*, Princeton University Press, Princeton, 2001 <https://www.jstor.org/stable/j.ctt7s610>

Jahoda G. : "Beyond Stereotypes," *Culture and Psychology*, Vol. 7, 2007 pp. 181–197.

Kant E. *Projet de paix perpétuelle* 1795 Paris, éditions Nathan les intégrales de Philo 2001.

Kaufmann J. C. *La chaleur du foyer: analyse du repli domestique* Paris, éditions Kincksieck, 1988.

Khodyakov D. "Trust as a process: A Three-dimensional Approach" in *Sociology*, London, 2007, vol. 41, pp. 115–132.

Kreps D. M. "Corporate Culture and Economic Theory" in Alt J. and Shepsle K. *Perspective on Positive Political Economy*, University Cambridge Press, 1990.

Kydd A. : *Trust and Mistrust in International Relations*, Princeton University Press 2005 <https://doi.org/10.2307/j.ctv39x4z5>

Lagroye J., *Sociologie politique*, 1997.

Laufer R. and Orillard M (dir), *La confiance en question*, Paris, édition Harmattan, collection logiques sociales, 2000, 407 p.

Lazuech G. *Toute confiance est d'une certaine manière une confiance aveugle*, Mayenne, éditions Plein feux, 2002, 43 p.

Lebow R. N. : "The role of Trust in International Relations," *Global Asia*, Vol. 8, n°3, September 2013.

Luhmann, N. *Trust and Power: Two Works by Niklas Luhmann*. 1979, Translation of German originals Vertrauen 1968 and Macht 1975. Chichester: John Wiley.

Luhmann N. "Familiarity, confidence, Trust: problems and alternatives" in Gambetta D. *Trust making and breaking Cooperative relations*, Oxford, Blackwell editions, 1988.

Luhmann N. *La confiance: un mécanisme de réduction de la complexité sociale*, Paris, édition Economica 2006, 123 p.

Mangematin V., Thuderoz C., *Des mondes de confiance: un concept à l'*épreuve de la réalité sociale, Paris, éditions CNRS, 2003, 295 p.

Marty and Appleby, *Fundamentalism and the state: remaking politics, economics and militancy*, Chicago Press of University, 1991.

Michel, T. Time to get emotional: Phronetic reflections on the concept of trust in International Relations. *European Journal of International Relations*, *19*(4), 869–890, 2013 <https://doi.org/10.1177/1354066111428972>

Michel, T. Trust and International Relations. In P. James, K. Fierke, A. Freyberg-Inan, S. Gartner, S. Lobell, & N. Sandal (Eds.), Oxford Bibliographies in International Relations Oxford University Press, 2016 https://doi.org/10.1093/obo/9780199743292-0192

Misztal B. A. *Trust in modern societies* London, Cambridge, Polity Press, 1996.

Möllering G. *Trust, Reason, Routine, Reflexivity*, Oxford, Elsevier, 2006, 217 p.

Nyhan R. C. "Changing the paradigm Trust ad Its Role in Public Sector Organizations" in *The American Review of Public Administration* Vol. 30, March 2000 pp. 67–105.

Ostrom E. and Walker J. *Trust and reciprocity*, New York, Russel Sage Foundation, trust collection, volume 6, 2003, 409 p.

Pettigrew Pierre S. *Pour une politique de la confiance* Québec Les éditions du Boréal, 1999.

Putnam R., Leonardi R and Nanetti R. *Making democracy work: civic tradition in modern Italy*, Princeton, Princeton University Press, 1993.

Putnam R. D. "The strange disappearance of civic America" in *American Prospect*, n° 24, 1996, pp. 34–49.

Quere Louis, *La confiance*, Paris, Hermés-Sciences, *Réseaux* (19, 108), 2001, 240 p.

Quere Louis et Ogien A. (Dir) *Les moments de la confiance. Connaissance, affects et engagements*, Paris, Economica, Coll études sociologiques, 2006, 232 p.

Ruscio K. P. "Jay's Pirouette, or Why Political Trust is not the Same as Personal Trust" in *Administration and Society*, November 5, 1999, n°31 Vol. 5.

Russel; Hardin (Dir), *Distrust*, New York, Russel Sage foundation, collection on Trust, volume 8 1992, 334 p.

Russel, Hardin (Dir), *Trust and* New York, *Trustworthiness*, éditions Russel Sage foundation, collection on Trust, volume 4, 2002, 233 p.

Sa'adah A. "Regime Change: Lessons from Germany on justice, Institution building and democracy," in *Journal of conflict resolution*, n° 50, volume 3, June 2006.

S&P Top Geopolitical Risks of 2024, 2024, <https://www.spglobal.com/en/research-insights/market-insights/geopolitical-risk>

Scott K. D. "The causal relationship between Trust and the assessed value of management by objectives", in *Journal of management*, vol. 6, 1980, pp. 157–175.

Seligman A. B. *The problem of Trust*, Princeton University Press edition, 1997, 232 p.

Seligman A. B. "Trust and civil society" in Tonkiss F.; Passey A., "*Trust and civil society*," London, Macmillan Press, 2000, pp. 12–30.

Simmel G., *Conflict and the web of group affiliations*, New York, The Free Press, first edition 1955.

Simmel G *Sociologie et épistémologie*, Paris, Presses Universitaires de France, 1991 (coll. Sociologies).

Spencer S. J., Fein S., Wolfe C. W., Fong C., Dunn M. "Automatic Activation of Stereotypes: The Role of Self Image Threat", *Personality and Social Psychology Bulletin*, Vol. 24 n° November 11, 1998.

Sztompka P. *Trust a sociological theory*, Cambridge University Press, 1999, 211 p.

Tilly C. *Trust and rule*, Cambridge University Press, 2005, 197 p.

Thuderoz C., Mangematin V. and Harrisson D. *La confiance: approches économiques et sociologiques*, Paris, édition Gaëtan Morin, 1999, 325 p.

Thuderoz C. "La confiance en questions" in Mangematin V., Thuderoz C. Ouvrage collectif, *Des mondes de confiance: un concept à l'*épreuve de la réalité sociale. Paris, éditions CNRS, 2003, pp. 19–30.

Tonkiss F.; Passey A. Collective work, *Trust and civil society*, London, Macmillan Press, 2000, 190 p.

Touraine Alain, *Production de la société*, Paris, Seuil, 1973.

Tyar A. F. *Les aléas de la confiance: gouverner, éduquer, psychanalyser*, Paris, édition l'Harmattan, collection psychanalyse et civilisations, 1998, 330 p.

Van de Wetering C. "Mistrust among democracies: Constructing US-India insecurity during the Cold War" in Haukkala H., van de Wetering C., Vuorelma J.in *Trust in International Relations, Rationalist, Constructivist, and Psychological Approaches* Routledge 2018.

Vuorelma J. "Trust as a narrative: Representing Turkey in Western foreign policy analysis" Haukkala H., van de Wetering C., Vuorelma J. in *Trust in International Relations, Rationalist, Constructivist, and Psychological Approaches* Routledge, 2018.

Warren M. E. (Dir) Collective work "*Democracy and trust*" United States, Cambridge University Press, 1999, 370 p.

Wheeler N.: *Trust Building in international relations*, Peace Prints, 2011 <https://wiscomp.org/peaceprints/4-2/4.2.9.pdf>

Wille T. and Martill B. : Trust and calculation in international negotiations: how trust was lost after Brexit, *International Affairs*, Volume 99, Issue 6, November 2023, pp. 2405–2422, <https://doi.org/10.1093/ia/iiad243>

World Economic Forum (WEF a): 5 things you need to know about geopolitics in a fractured world, <https://www.weforum.org/agenda/2024/08/geopolitics-democracy-trade-misinformation>

World Economic Forum (WEF b) The Global Cooperation Barometer 2024, insight report <https://www.weforum.org/publications/the-global-cooperation-barometer-2024/>

CHAPTER 1

Democracy and Trust

Imagine a world where free elections gave way to abuse of power and the end of freedoms, while citizens remained apathetic.

The late 1990s marked a pivotal moment in global history, characterized by the collapse of the communist bloc, the opening of societies to democratic governance, and the widespread adoption of market economies. These changes fueled a wave of optimism, sparking hopes for the expansion of civil liberties and democratic ideals across the globe. Some authors even mentioned that this could be "The end of history" (Fukuyama, 2005). In the following decades, pro-democracy movements emerged with renewed vigor in diverse regions, from the Arab Spring in the MENA to the "occupy central" protests in Hong Kong (2014), but also in Georgia, Iran, and Belarus. Yet, despite their fervor and global attention, these movements largely failed to establish enduring democratic systems.

Today, the landscape is further complicated by the rapid evolution of technology, which increasingly undermines democratic principles and the rule of law. Simultaneously, geopolitical tensions and armed conflicts exacerbate these challenges. Russia's war of aggression on Ukraine in February 2022 and the intensifying violence in the Middle East since 2023 are stark reminders of the fragility of peace and the growing threats to human rights and freedoms. These developments raise pressing questions about the resilience of democratic systems.

Western democracies, once seen as bastions of stability, now grapple with a profound crisis of trust. Polarized narratives, disinformation, and the rise of populist movements have called into question the foundational principles of democratic governance. The recent election of Donald Trump as the 47th President of the United States, along with his administration's appointments, raises questions about the future of American democracy. These concerns include the rule of law, civil liberties, and the independence of both the media and judiciary—institutions that have helped make the United States one of the world's leading democratic models in recent decades. Are we witnessing a global democratic decline? Is this primarily a crisis of institutional trust, or does it reflect deeper systemic vulnerabilities? And perhaps most critically, could the current trajectory jeopardize both free speech and public confidence in democratic institutions? These are urgent questions for a world increasingly defined by uncertainty and division.

In times of peace and stability, trust is a fundamental condition for a fair and cooperative society. Trust plays a vital role in democracies by legitimizing institutions and rules, which are enforced not only through power but through mechanisms of checks and balances. It serves as a mechanism to reduce complexity in society. Systems and institutions rely on confidence and trust for the acceptance of regulations and justice, the implementation of social programs, and the achievement of stability. The alternative is a vicious cycle of distrust, leading to negative consequences in the economy, conflicts, and failed governance.

I. The Paradoxes of Trust and Democracy

Democracy is defined as a system of government with checks and balances, freedom of speech and assembly, and institutions through which citizens can express their preferences. It includes constraints on executive power through the rule of law and inclusive universal suffrage for electing national leaders (Dorenspleet 2004). The balance of power between different parties and institutions serves as one of democracy's key regulatory functions. Democratic institutions also ensure freedom of speech, independent media and justice systems, civic rights, and individual freedoms.

Democracies create a balance between rights and obligations, a system of check and balances reinforced by the rule of law. The rule of law is one of the conditions of democracy: it organizes political competition (by making it more secure) and reduces attempts at arbitrary power (Sa adah 2006:305). But what would happen if these principles were eroded by progressive abuse of power? What if democratic institutions became paralyzed or deteriorated?

Public institutions secure citizens' basic rights and freedoms. When effective, these institutions foster a sense of security and build trust among citizens in their structures and procedures (Rawls 1996). Hardin, who defines trust as an *"encapsulated interest,"* reaches this same conclusion through rational analysis. Since individuals naturally act in their own self-interest, they are naturally untrustworthy for society. This is why laws and conventions are needed to protect the general interest (Hardin et al. 1996). But what happens when governments, leaders or an oligarchy do not respect the laws?

Trust is a prerequisite for democracy. According to Charles Tilly, democracy requires more trust than any other political system, as it establishes relationships of reciprocal trust between decision-makers and citizens, as well as between ruling and opposition parties (Tilly 2005). Many scholars argue that trust enhances the legitimacy and effectiveness of democratic governments (Braithwhite et al. 1998, Hetherington 1998). This is particularly crucial for new governments taking over from untrustworthy predecessors (Dogan et al. 1998). While research has extensively examined trust in democracy (Stztompka 1999, Putnal et al. 1993, Warren 1990, Tilly 2005), there is limited investigation into how trust and democracy intersect with digitization, particularly regarding new technologies and their impact on public perceptions. This chapter will not attempt to address all these aspects but instead offers preliminary ideas that could guide future research.

Trust connects ordinary citizens with the institutions that represent them (Bianco et al. 1994) Liberal democracy ensures that citizens place their trust in government but can also withdraw it. Trust is a communion of interests between those who govern and those who are governed (Ruscio, 1999).

There are two approaches to the question of trust in institutions: the cultural approach and the rational approach. For the so-called cultural theory, trust in institutions is an extension of interpersonal trust. It is external to the institution (Almond et al. 1963, Putnam 1993, Foley et al. 1999). In the rational approach theory, trust in institutions depends on the reliability of institutions and on their institutional performance (Dasgupta 1988). Trust is therefore a consequence, not a cause, of institutional performance. (Almond et al. 1963, Putnam 1993, Levi 1996, Inglehart 1997, Foley et al. 1999…). Can we truly consider institutions in modern societies without any subjective viewpoints? Moreover, is there a way to evaluate institutional performance beyond purely technocratic standards?

As a matter of fact, it is important to mention that democracy is not equivalent to trust. Liberalism and liberal democracy emerged in Europe out of distrust of the traditional political representatives constituted by the nobility and the clergy. This is where the question of trust in democracy is paradoxical. Democracy can be related to both trust and distrust, while at the same time it remains vulnerable when distrust reaches certain peaks and levels.

More democracy has meant more knowledge (transparency) and less trust in the authorities (Dunn 2000, Sztompka 1999). Political situations are marked by problems, conflicts of interest or identity in which (political) parties can provide support and keys to understanding to help resolve dilemmas (Warren 1990). Democratic mechanisms enable individuals to question assumed trust, limit the power of politicians (through voting in particular) and thus limit the potential damage to themselves (Warren 1990:2).

Democracy excludes absolute trust, though this does not mean democracy can function without trust (Warren 1990). Democracy operates on a limited trust that is guided by critical thinking. At the same time, it requires a broader trust in the system itself, even while maintaining a critical perspective. This foundation of trust is also essential for individuals to successfully organize collective action.

As societies grow more complex and interconnected, people experience greater diversity, mobility, choices, and online connections. This interdependence makes individuals more vulnerable, since they have less control over their environment and social systems (Offe 1996). The rise of social media has heightened these vulnerabilities, as interactions are influenced by algorithms and online communities are hosted by profit-driven social media platforms. What if online social media platforms could influence elections, polarize narratives and lead to apathy?

According to Barber, trust manifests in the expectations that social actors hold toward one another. These expectations stem from the belief that social interactions sustain and fulfill what is seen as a natural moral and social order (Luhmann 1979, Barber 1983). Trust simplifies life for individuals by providing security and allowing them to rely confidently on their systems and institutions. From a functional perspective, trust and democracy serve as complementary tools for fostering collective action and cooperation among stakeholders.

Democracies are exposed to regular crises. **Terror Attacks** are extremely impactful as they create fear and impact trust in public institutions. The 09/11 attacks in the United States in 2001 changed the world. Less spectacular terror attacks such as the ones in Paris in November 2015 and in Brussels in March 2016 also profoundly impacted trust, both at the social and institutional levels, while reshaping the security landscape. When violent acts target public spaces, they erode citizens' sense of safety and confidence in the state's ability to protect them, fueling fear and division. Trust in governments may waver if responses are seen as inadequate or disproportionately restrictive, raising concerns about civil liberties. Simultaneously, these events drive shifts in security policies, leading to increased surveillance, tighter border controls, and expanded counterterrorism measures. Such responses, while aimed at preventing future attacks, can also spark debates about privacy, human rights, and social cohesion. Moreover, terror incidents often amplify distrust between communities, particularly when they are exploited for political agendas or misinformation campaigns. As a result, societies face the dual challenge of strengthening security while preserving the fundamental values of openness, inclusion, and democratic resilience.

The more complex and differentiated a society, the less likely it is that interactions will spontaneously be based on trust (Hardin 2004). Crises of democracy are punctuated by crises of trust, and sometimes by social movements which question the reason of being and the fundamental grounds of democracy.

II. Democracy Regressing Worldwide

The late 1990s marked a transformative period in global history, with the collapse of the Eastern Bloc, the opening of societies to democracy, and the widespread adoption of market economies. These events sparked new hopes for civil societies around the world, suggesting a future where democratic principles and economic integration could thrive. The Rose Revolution in Georgia (2003) and the Orange Revolution in Ukraine (2004) followed by the Euromaidan (occupy Maidan) movement in 2014, both popular uprisings, resulted in government changes. By the 2010ies, waves of pro-democracy movements also gained momentum in various parts of the globe, from the Arab Springs in the MENA to the "occupy central" uprisings in Hong Kong but also spreading to Iran, and Belarus. However, despite their intensity and the global attention they garnered, few of these movements succeeded in establishing sustainable democratic systems capable of enduring overtime.

Later on, the emergence of unexpected and destabilizing armed conflicts further undermined these democratic aspirations. The war of aggression against Ukraine, which began in February 2022, and the escalation of violence in the Middle East since 2023 have raised grave concerns about human rights violations and the erosion of international norms. Adding to these challenges is the rise of advanced technologies, which increasingly threaten democratic principles and the rule of law by enabling disinformation, surveillance, and the manipulation of public opinion.

Western democracies, long regarded as global exemplars of governance, now face their own crises of trust and legitimacy. Polarized narratives, widespread disinformation, and the rise of populist movements have eroded public confidence in democratic institutions. This erosion is exemplified by the recent election of Donald Trump as the 47th President of the United States and the subsequent appointments within his administration, which have sparked widespread debate about the future trajectory of American democracy. Concerns have been raised over the preservation of the rule of law, civil liberties, and the independence of critical institutions such as the media and the judiciary. Specifically, the amnesty granted to those who stormed the Capitol on January 6, 2021, raises serious concerns about the future. The signing of drastic executive orders to cut the federal budget, the internment of undocumented migrants at Guantanamo, and attacks targeting transgender communities demonstrate a level of brutality that is hardly compatible with the rule of law, as they disregard parliamentary or civil opposition and create legal uncertainties.

Elon Musk's recruitment of young computer scientists to collect—or even hack—federal budget data was fortunately slowed by a court ruling, but one could argue that the damage has already been done. It was followed by, the publication of often erroneous data on the X account of the DOGE department for federal spending reductions before Elon Musk finally left the Trump Administration.

Trust is no longer there; foreign propaganda narratives seem almost mild compared to a rhetoric that actively undermines confidence in the country's institutions and seeks to convince citizens that public funds have been recklessly squandered on absurd causes.

Are we now witnessing a major step backward in the progress of democracy? Is this a symptom of a worldwide democratic decline, or are these challenges indicative of a deeper transformation in how governance and trust are negotiated in the twenty-first century? These questions resonate profoundly as democratic systems are increasingly tested by internal and external pressures.

According to the 2024 V-M democracy index, less than 13 percent of the world's inhabitants live in a liberal democracy that respects fundamental rights and freedoms, and 72 percent of the world's population lives in an autocratic state. The index of "real" democracy seems to be under pressure, particularly in Europe, but also in India and across the Atlantic. Human rights, press freedom, and the independence of the judiciary remain under threat in many countries, with an increase in disinformation, populism, and outside influences. The birth of the Arab Springs has given up the ghost in the countries south of the Mediterranean, while the "democratic" countries of the Middle East, such as Israel and Lebanon, are seeing their systems challenged and called into question.

More frighteningly, the rise of totalitarian models, encouraged by large powers such as Russia, raises questions about the massive use of surveillance methods on the population. This chapter will take as its starting point concrete examples of the retreat of democracy and human rights, to highlight the understanding of new models and tools that oppose them head-on and fundamentally ...

The decline of democracy on a global scale seems self-evident, but what about the analysis and the figures? According to the 2024 V-Dem report, which is based on an analysis of 60 indices and 500 indicators, it is possible to distinguish four main categories of regime in terms of fundamental rights

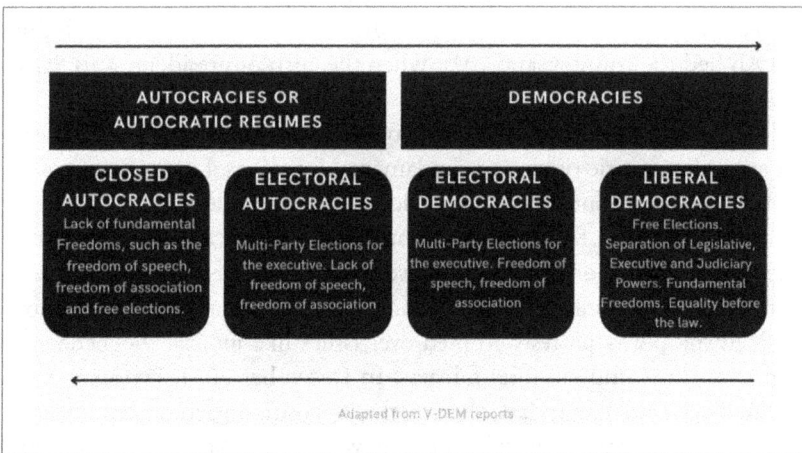

AUTOCRACIES OR AUTOCRATIC REGIMES		DEMOCRACIES	
CLOSED AUTOCRACIES	ELECTORAL AUTOCRACIES	ELECTORAL DEMOCRACIES	LIBERAL DEMOCRACIES
Lack of fundamental Freedoms, such as the freedom of speech, freedom of association and free elections.	Multi-Party Elections for the executive. Lack of freedom of speech, freedom of association	Multi-Party Elections for the executive. Freedom of speech, freedom of association	Free Elections. Separation of Legislative, Executive and Judiciary Powers. Fundamental Freedoms. Equality before the law.

Adapted from V-DEM reports

Figure 1. Democracies and autocracies

and the balance of power: closed autocratic states, electoral autocracies, electoral democracies, and liberal democracies—see figure 1.

According to the 2024 report, which does not take into account a number of subsequent coups d'état in Western Africa, 72 percent of the world's population, that is, 5.7 billion people, live in autocracies, whether closed or electoral. This represents a 46 percent increase on the previous decade. Closed autocracies are represented by countries that have been singled out for their human rights record, such as North Korea, China and Iran. Electoral autocracies include a variety of regimes such as Russia and Turkey. V-Dem also considers India to belong to this group. Electoral democracies would represent 58 countries with a total of 16 percent of the population, and liberal democracies 13 percent, with possible attacks on institutions, as in the USA after the assault on the Capitol, or in Brazil with the insurge of extreme rights movements after the election of President Lula Da Silva in the end of 2022 and beginning of 2023..

In terms of geographical representation, the MENA—Middle East-North Africa—region is predominantly autocratic, with 98 percent of its population living in autocracies, and the remaining 2 percent in Israel, a country that is nevertheless subject to human rights violations—and attacks on democratic institutions which started before the deadly terror attack on October 7th 2023 and the war crimes in Gaza as recognized by the International Criminal Court (ICC). It should also be noted that Tunisia enjoyed a sometimes-turbulent ten-year period of democracy between 2011, when President Zine el-Abidine Ben Ali fled the country, and 2021, when the current president, Kaïs Saïed, took full powers.

The Asia-Pacific region is the most populous, yet nine out of ten people do not enjoy democratic rights and fundamental freedoms. Liberal democracies include Japan and South Korea, while electoral democracies include Indonesia, Mongolia and Nepal. South Korea recently faced a significant democratic crisis when allegations of corruption and abuse of power emerged within the government, intensifying political polarization. Public trust in political institutions declined sharply as protests erupted over issues like judicial independence, media freedom, and electoral fairness. In December 2024, President Yoon Suk Yeol declared martial law, alleging that opposition parties were engaging in "anti-state activities" and collaborating with "North Korean communists." This unprecedented move led to widespread protests and political turmoil.

The National Assembly swiftly convened, voting unanimously to annul the martial law decree, which was subsequently lifted by President Yoon. Following these events, Yoon was impeached and suspended from office, pending a Constitutional Court decision on his removal. In January 2025, he was arrested on charges of insurrection, marking the first time a sitting South Korean president faced such charges. This series of events has raised profound concerns about the state of democracy in South Korea and its political stability. While South Korea remains a strong democracy in many respects, these challenges have raised concerns about its resilience and the ability of its democratic institutions to uphold accountability and public confidence.

Sub-Saharan Africa is home to a number of democracies, including South Africa, Ghana and the Seychelles. Many countries have had autocrats in power for decades, such as Cameroon and Equatorial Guinea, while others are experiencing devastating conflicts, notably in the Horn of Africa. Successive coups d'état in West Africa do not augur a democratic opening, but rather a resurgence of conflicts.

Russia, Belarus and the Central Asian republics offer a panorama of electoral or closed autocracies, while Eastern Europe and the Caucasus are mainly home to electoral democracies, with populist governments in several countries such as Viktor Orbán's Fidesz Hungary, or the concerning rise of extreme rights movements in the Netherlands, Austria, Belgium France and Germany in 2024 and 2025.

Romania faced a democratic crisis in December 2024 when its Constitutional Court annulled the first round of presidential elections. The court cited clear evidence of foreign interference through a coordinated disinformation campaign, allegedly led by Russian actors. The interference spread mainly through social media, where TikTok played a key role in boosting far-right messages and raising the profile of candidate Călin Georgescu. In response, the European Commission launched an investigation into TikTok, examining whether the platform broke the Digital Services Act by failing to address risks of interference during elections. This case reveals how vulnerable democratic systems are to disinformation and outside manipulation, showing why strong protection is needed to keep public faith in electoral integrity. More optimistically, only 12 percent of Latin American residents live in autocracies, and these are relatively medium-sized countries like Cuba, El Salvador, Nicaragua and Venezuela. The 2024 elections in Venezuela show however that the opposition to President Maduro is still very active and enjoys a huge popularity. María Corina Machado and Edmundo González Urrutia who officially won the elections, but which victory was denied, were awarded the 2024 Sakharov Prize of the European Parliament.

One argument often put forward is that democracy is just another power model "imposed" by the West. The number of pro-democracy movements firmly repressed all over the world, from Hong Kong to Teheran, however, strongly opposes the idea that only Westerners love freedom and respect for human rights. But then, if we don't look at multi-party elections, what about other fundamental freedoms? The rule of law and the integrity of elections are deteriorating in many countries. According to V-Dem, by 2024, global democratic standards have regressed to levels last seen in 1986, prior to the collapse of the Soviet Union and the fall of the Iron Curtain in Europe.

Faces of Repression and Repression Without Faces

In **Russia**, the crackdown on dissent has reached alarming levels, particularly since the death of opposition leader **Alexei Navalny** in 2024. His death in an Arctic prison colony marked a turning point, sparking intensified persecution of his supporters and other Kremlin opponents. Human rights organizations reported a surge in political prisoners by December 2024, with individuals imprisoned merely for expressing anti-war views or criticizing government policies.

Additionally, disturbing reports reveal that that **Ukrainian prisoners of war** held by Russian forces are being subjected to **ill-treatment**. These practices, which constitute human rights violations and **war crimes**, have been strongly condemned by international organizations.[7,8]

In **Iran**, the crackdown on dissenting voices and minorities has intensified alarmingly since the nationwide uprising sparked by **Mahsa Amini's** death in September 2022. Arrested by the **morality police** for allegedly wearing her headscarf incorrectly, she died in custody. Her death ignited a widespread protest movement—known as Woman, Life, Freedom—led primarily by Iranian women demanding freedom and equality.

The Iranian regime responded to these protests with brutal repression, including thousands of arrests, summary trials, death sentences, and systematic violence against protesters. In **2024**, Iran carried out the highest number

---

[7] <www.wsj.org.>
[8] <www.amnesty.org>

of executions in three decades. These executions, following unfair trials and forced confessions, are typically based on vague charges. Despite their own imprisonment, figures of the Iranian resistance, including 2023 Nobel Peace Prize laureate Narges Mohammadi, continue to denounce this repression.[9]

In **China**, repression takes a distinctive form—faceless and systematic, rendering opponents anonymous. People vanish for periods ranging from weeks to indefinite durations. This repression is woven into a system of **mass surveillance and social control**. The government monitors citizens closely, censors information, and suppresses opposition. The exact number of political prisoners remains impossible to document.

Mistreatment throughout Chinese detention centers remains a reality.

Many detainees, including political prisoners, activists, ethnic minorities such as the Uyghurs, and religious minorities such as members of the peaceful Falun Gong movement, are targeted by large-scale repression. These practices have been criticized by international human rights organizations, which denounce repression legitimized under the guise of national security and party stability.

Despite differences in method, the outcome is the same: in Russia, Iran, and China, repression silences opposition, obliterates individual freedoms, and serves as the backbone of regimes built on fear.

Here, too, we seem to be witnessing a mass retreat from certain fundamental freedoms on a global scale, which is all the more astounding given that new technologies and the internet allow information to circulate at mind-boggling speed. In this context, how is it possible that freedom of expression has deteriorated in 35 countries in 2022 (V-Dem 2023). In this age of social networking and multimedia? Media censorship is deteriorating in 37 countries. In almost 40 countries, governments have increased their control over civil society organizations with increasingly restrictive rules; while in 37 countries, a state of repression of civil society is looming.

Press freedom is also under threat in many countries. The Reporters Without Borders (RSF) 2025 ranking reveals a worsening of press conditions, due to political, social and technological instabilities, and at a time when it is becoming almost common place for journalists to be murdered in many parts of the world, starting with areas in conflict such as Gaza which has been

---

[9] <www.lemonde.fr.>

called "a graveyard for journalists." Conditions for the exercise of journalism are poor in seven out of ten countries, and satisfactory in only three out of ten. Among the countries ranked at the very bottom of the list are Iran 176[e], China 178[e] , where the number of imprisoned journalists is the highest in the world, and with the export of massive propaganda, and North Korea at number 179[e]. The report also denounces the context of insecurity and repression in (but not restricted to) Sub-Saharan Africa, the MENA region, and Latin America, with concerning scores for Saudi Arabia 162[e] just ahead of the Arab Emirates 164[e], Cuba 165[e], Belarus 166[e], Burma 169[e], Russia 170[e], Vietnam 173[e], President Isaias Afwerki's Eritrea 180[e], and Syria under Bashar el Assad's regime 177[e].

More than an opposition between autocracy and democracy, what we are witnessing is a retreat from fundamental freedoms and democratic principles as a model, but who is really aware of the danger this represents for democracy and freedoms around the world? How can we rely on autocrats to resolve global issues ranging from climate change to peace and security? Are we really to believe that our stable institutions in Europe and North America will protect us for much longer from the propaganda of lawless states and the spread of freedom-destroying models? Who has an interest in what? And which states are willing and able to export disinformation to weaken Western democracies?

Disinformation and the pretense industry are prevailing. RSF's 2024 world Press Freedom Report notes that 118 countries and political players are involved in massive disinformation and propaganda campaigns, on a regular and systematic basis. The difference between the true and the false, the real and the artificial, facts and artifacts is becoming blurred, jeopardizing the right to information. The proliferation of large-scale manipulation tools is a cause for concern. We will revisit this topic in the chapter on disinformation.

III. Trust in Western Democracies at Stake

Western democracies have been confronted with crisis that is endangering their essence. The emergence of populist movements and their success in the polls are an evident sign of their vulnerability. The yearly Edelman Trust report reveals a general decline in trust in systems. Media reports indicate not a fundamental rejection of democracy itself, but rather widespread attempts to undermine democratic principles and a major crisis of trust across several countries. The next chapters will explain how democracies have entered a global image war, marked by hate speeches and populist ideas.

Democracies are increasingly challenged by narratives that exploit deep-seated fears and prejudices, fracturing societies along political, ethnic, gender, and religious lines (Moravcsik 1997, Bollman 2022) and to amplify scapego-ating approaches to create fear and anger. From the rise of Euroscepticism, which culminated in Brexit, to the divisive rhetoric surrounding migration, gender, and religion, these narratives have often proven effective in manip-ulating public opinion and creating sharp societal divides. Identifying these polarizing narratives is crucial because they undermine social cohesion, fuel extremism, and threaten democratic stability. By recognizing and understand-ing these narratives, society can better counteract their harmful effects and prevent the spread of disinformation.

As an example, the success of Brexit was largely driven by emotionally charged and divisive narratives that polarized the British public, creating strong "us vs. them" dynamics (Cervi et al. 2019: 134). The Brexit campaign also employed misleading information, such as the infamous "Brexit bus," which falsely claimed that leaving the European Union (EU) would result in more funding for the National Health System (NHS). Other central themes of the campaign included the notion that Britain played a significant role on the global stage and "did not need Europe," alongside the idea of "taking back control" and empowering the people. At the time these narratives were very successful and won the referendum to leave the EU, but the outcome of Brexit overall lowered Euroscepticism and mostly halted similar movements in other European countries.

Since the 2010s, Viktor Orbán's political regime in Hungary has under-mined the independence of both the media and the judiciary. The European Parliament has raised persistent concerns about democracy and fundamental rights in the country. Key issues include the functioning of the constitutional and electoral system, judicial independence, corruption, conflicts of interest, and freedom of expression—particularly media pluralism. Additionally, academic freedom, freedom of religion, freedom of association, and the right to equal treatment have been identified as areas of concern (European Parliament, 2022).

As we have seen earlier, Romania's recent presidential elections faced controversy after the Constitutional Court annulled the first round, citing significant foreign interference, including a coordinated disinformation campaign linked to Russian actors. Social media platforms, particularly TikTok, were instrumental in spreading far-right narratives and boosting certain candidates. This incident highlights the growing threat of disinformation to democratic processes and raises concerns about the integrity of elections in the digital age.

In the United States, measures restricting press freedom have been reported by the organization *Associated Press* in February 2025 and appear to be in direct contradiction with the First Amendment of the U.S. Constitution.

Recent Concerns in Europe and in the United States

In **Europe**, the assassination of investigative journalist Daphne Caruana Galizia on October 16, 2017, in Malta underscored the perils faced by those exposing corruption and misconduct. Her murder, carried out via a car bomb, highlighted the vulnerabilities of journalists even within established democracies and the lengths to which corrupt entities will go to silence scrutiny. This tragic event sparked widespread international condemnation and prompted calls for stronger protection for journalists and greater accountability for those who threaten press freedom.

Unfortunately, political pressures on independent media and civil society are also increasing across Europe. In countries like Hungary, the government has tightened control over the press, systematically dismantling independent news outlets and using state resources to influence public discourse. Poland, until recently, faced similar restrictions on judicial independence and media freedom. These trends raise concerns about democratic backsliding within the EU, where press freedoms—once considered secure—are now under growing political and economic pressure.

In the **United States**, a **new form of censorship** is emerging, particularly concerning discussions on **climate change and biodiversity**. Political polarization has led to concerted efforts to suppress scientific research and discourse on environmental issues. This includes attempts to restrict educational content, manipulate public narratives, and undermine policies aimed at addressing ecological crises. Certain states have banned discussions of climate risks in schools, while corporate interests and lobby groups exert pressure to downplay the urgency of climate action. This form of ideological suppression not only limits public awareness and understanding but also poses severe risks to environmental sustainability and informed policy-making. The restriction of press freedoms and the suppression of federal archives are also alarming.

The scapegoating approach is also recognized as one of the strategies of manipulation to polarize the political debate (Deutsch 1958, Girard 1986, Hersh 2013, Goodhart 2017, Betz et al. 2021, Bauer et al. 2023).

Narratives about migration and ethnic minorities often exploit public fears by spreading disinformation that links migration to crime, unemployment, inequality, and other societal challenges. These narratives subtly frame migration in a negative light, deepening polarization on the issue.

In many European countries, racism and xenophobia have played a significant role in extreme nationalist movements, reinforcing divisions between groups based on racial and ethnic lines. Central to these narratives is the "myth of a homogeneous nation," which not only fuels hostility toward ethnic minorities and pluralistic societies but also targets liberals who advocate for diversity. This highlights how such narratives create division—not only between the far right and ethnic minorities but also across broader political lines concerning migration.

In January 2025, President Donald Trump implemented a series of highly controversial immigration policies on the other side of the Atlantic. Declaring a national emergency at the U.S.-Mexico border, he deployed troops to strengthen border security and authorized the immediate deportation of individuals entering the country illegally. Additionally, he signed an executive order seeking to end birthright citizenship for children born in the United States to undocumented immigrants, a move that has sparked fierce legal challenges and debates over its constitutionality. These measures have reignited discussions about the balance between national security, human rights, and the broader social and economic implications of such aggressive immigration policies.

Polarizing narratives are highly adaptable to the political landscape. During the COVID-19 pandemic, for example, those promoting divisive rhetoric about migrants quickly shifted their focus to health concerns. They falsely depicted migrants as less likely to get vaccinated, comply with restrictions, or as contributing to the virus's spread.

The World Health Organization (WHO) Has emphasized that COVID-related disinformation fueled xenophobia, intensified public polarization, and deepened societal divisions (WHO, 2020). Research indicates that narratives linking migration to health and economic issues proved even more effective at polarizing society than identity-based narratives, such as the "myth of a homogeneous nation." These discourses often translate into concrete actions and policies.

Belief in narratives of ethnic supremacy represents an extreme form of racism, but subtler forms of racism and xenophobia also deserve attention, beyond overt acts of violence. Discrimination against ethnic minorities can take many forms, such as being excluded from job opportunities simply because of a foreign-sounding name, making employment more difficult to secure.

Migrants and refugees also face significant barriers to education. In Poland, for example, reception centers are sometimes located far from urban areas, limiting access to schooling for refugee and migrant children. This cycle of discrimination is further reinforced by xenophobic narratives that mis-interpret such disparities as justification for their views. These narratives become self-perpetuating, contributing to real social inequalities, including higher unemployment rates, lower educational attainment, and increased homelessness among foreigners and ethnic minorities.

Furthermore, looking toward the future, our racial biases could have implications for the development of AI. Algorithmic learning can absorb human biases, potentially reinforcing illegal ethnic discrimination and lead-ing to what has been termed the "automation of discrimination" (Zuiderveen Borgesius, 2018, pp. 11–12). Illustrating the consequences of discriminating narratives, and how it might continue to divide societies.

Finally, the exploitation of divisive narratives during political campaigns serves as a calculated strategy to systematically discredit and undermine public institutions, including the judiciary, media, and electoral systems, by persistently questioning the fundamental fairness and legitimacy of established institutions and governments. This erosion of trust is accom-plished through carefully crafted messaging that targets the credibility of these pillars of democracy. Donald Trump's 2024 campaign in the United States has deliberately and systematically fueled anger and deep-seated resentment against political opponents by methodically spreading false and inflammatory information about migrants. These targeted disinformation campaigns serve to create division and distrust within the electorate. The violent assault on the Capitol in 2021, followed by the prolonged and systematic questioning of the 2020 election results, have not only directly jeopardized the stability of the U.S. democratic system but have also exposed its heightened vulnerability to coordinated misinformation campaigns that are unfolding now. The cumulative effect of this sophisticated and calculated use of divisive and hate narratives goes beyond mere political rhetoric—it systematically undermines public confidence in democratic institutions, weakens social cohesion, and threatens the very foundations upon which democratic societies are built. But let's come back to the origins and to the theory of trust. How does it work?

IV. Individual or Social Trust? Analyzing How Trust Works in Democracies

For some authors, such as Francis Fukuyama, interrelationships are situated at the level of concentric circles of trust, extending from interpersonal trust to more abstract collective or social trust (Fukuyama 1995). The narrowest of these circles is trust between family members, sealed by relationships of intimacy and proximity. Next comes the trust of people we know personally, such as neighbors, friends and business partners. This circle also includes relationships of intimacy and proximity, followed by social trust. The last circle is made up of people with whom we feel we share common bonds, but who are mainly "anonymous others," not really known, but who make up a community in our imagination (our compatriots, members of our ethnic or religious group, our gender, generation or profession ...).

How can we distinguish or link individual and collective trust? In group analysis, it is sometimes common or tempting to distinguish between the micro-social level (interpersonal trust) and the macro-social level (social trust). Mark Warren, for example, distinguishes between generalized and limited trust within a group (Warren 1999). While the first level involves in particular local actions and face-to-face interactions, the macro-social level encompasses more global interactions between groups and individuals, nations and institutions. However, the idea that there is a micro-social level of trust concerning the individual; and a macro-social or institutional level of trust would be an oversimplification. There would be no antinomy between the psychological and the social, but rather complementarity marking the two sides of the same reality (Finc, 2005).

Trust, in the broadest sense of the term, that is, the ability to rely on one's own expectations, is a basic fact of life in society.

Between the level of individual trust and that of trust between large groups or communities on a national level, lie the relationships of trust between groups of more moderate size, such as within families, between families, in neighborhood relations, professional, religious, political and community relations and, last but not least, communities that act and interact on a national level.

There are not only two antinomic levels: the macro-social level and the micro-social level, which interact in places. Families, neighborhood groups, ethnic groups, regions, supporters of political groups etc. can also develop relations of mistrust due to their different interactions. Moreover, at the individual level, mistrust may also have spread to smaller groups (political parties, trade unions, associations), which may jeopardize stability.

To grasp this complexity, we first need to imagine a broad spectrum of relationships, taking into account individual relationships, those of small groups, and then those of large groups. Within the broad spectrum of relationships, we can find relationships between individuals, relationships between families, between trade unions, political parties, civil society players, and then interactions between large groups and wider communities. We can see the different circles of trust proposed by anthropologists, between local communities where trust reigns, and larger communities. The big difference, however, is that there is no optimal level of trust. A lack of trust can occur at any level of social relations.

Individuals are linked together by vast networks of relationships, and an individual may be connected to several groups. The sum of these relationships establishes patterns of trust/confidence between the groups themselves, whatever their size. In any relationship forming a network, there are opportunities for trust to break down as subjective feelings emerge. The image of a web woven by different networks of group affiliations relates to Georg Simmel's

Conflicts and the web of group affiliation, but it also refers to Charles Tilly's conception of trust in his book Trust and rule (Tilly 2005). For Tilly, there are networks of trust that explain the bonds of trust between restricted groups and institutions at sub-national and even transnational levels.

The methodological approach suggested here is to understand social and institutional (political) relations as different interactions within networks. But the difficulties of analysis cannot be reduced to this simple opposition.

We also need to understand how, if trust is a feeling based partly on cognitive (subjective emotions), this feeling can be attributed to groups. Talking about trust or distrust within groups presupposes that we can attribute attitudes, beliefs and emotions to groups. This is certainly a major stumbling block for the analysis of trust. Groups are not in themselves endowed with minds or the capacity to feel emotions, except through their members and their representations of themselves and those around them.

And even though the media sometimes claim that "so-and-so's community is moved by such-and-such an event," this is tantamount to essentializing the notion of community, and to saying in reality that the individuals who make it up felt the emotion, which was then expressed through their representatives. Groups by definition do not have an individual conscience, even though they deliberate, make decisions and behave as actors. Based on their actions, deeds and deliberations, however, we can recognize that these actions are not always motivated by explicit and easily identifiable considerations, but are based on a series of beliefs, attitudes and emotions shared by group members. It is therefore important to recognize that groups also have their own emotions, beliefs, symbols and collective imaginaries that influence their decisions. When it comes to relationships of trust and distrust, it's quite common to attribute these emotions to groups. In the media, for example, there are frequent references to crises of trust between states (South and North Korea, for example), between community groups, between belligerents in a conflict, but also between restricted groups and institutions (crisis of consumer confidence in the inflationary market).

Common sense therefore recognizes that relationships of trust exist between groups, although from a theoretical point of view it is difficult to confirm this. But if we recognize that relations of suspicion, hatred, revenge and revenge exist between groups, then we cannot rule out the idea that groups are also capable of demonstrating feelings of trust and compassion, or even gratitude and loyalty, as proposed by Georg Simmel (Simmel 1955).

Our work therefore proposes to recognize that there are positive and negative relationships between groups, based on subjective feelings, representations and behaviors attributed to the groups themselves, as well as to the individuals making them up (Haukkala et al. 2018). But it is not enough to assert that there are relations of trust and distrust between groups without taking the individual into account, because groups are made up of individuals. If groups act, interact and deliberate, or even fight, they do so only insofar as the individuals who make them up do the same. If we want to talk about groups experiencing trust or mistrust, we have to recognize that they only do so, because individuals (and not necessarily all individuals) making up the group also feel these same relationships of trust and mistrust, which are attenuated or amplified within the group by means of collective imaginary and this trend is even more vivid in the digital era with the rise of social media with the emergence of online communities. As noted by the authors Wille and Martill, "Most non-calculative conceptualizations of trust assume that the suspension of risk calculation takes place in the minds of the respective decision-makers." Such an understanding requires that the gap between mental states and collective action be bridged somehow and makes trust correspondingly difficult to assess. After all, in most cases, foreign policy decisions are not made by isolated individuals, but are produced collectively in political and bureaucratic organizations (Wille et al. 2023)

The only theoretical analysis that makes it possible to link these different theories while taking complexity into account is to recognize the weight of culture and collective feelings, whose representations are conveyed by the group. It is a constructivist approach that aims to avoid reifying concepts such as the state, the family, and the community, or drawing essentialist conclusions without analyzing complexities.

Starting from these different points, it is possible to recognize that relationships of trust exist between individuals, between groups, but also within groups, in a complex set of relationships that operate on a network basis. Acknowledging that groups can exhibit trust or mistrust does not mean confusing the group with the individual, nor mistaking the whole for the part and vice versa. When relationships between large groups show enough trust to enable cooperation, this does not mean that individual group members trust the other group. For example, in the context of European construction after World War 2, the national groups have established cooperation at European level based on trust, which does not mean that a Dutch automatically trusts a German and vice versa. It also means that, in a situation where a majority of individuals trust another group (distributive or horizontal trust), cooperation is possible and thus the emergence of trust between groups.

We must therefore avoid a methodological approach centered on the individual, which would attribute to the group the same feelings as to the individuals within it, just as we must avoid a methodological approach centered on the group (and more recently online communities), which would forget to take into account intra-personal and individual relationships within the group (the image of a network of interactions overcomes this problem). Finally, we must also avoid the pitfall of considering the group as an entity distinct from its members, as well as that of atomization (considering the individual without reference to these group affiliations and relationships including online communities) Indeed, feminist and communitarian theories have often criticized the failure to analyze the position of each individual as primus inter pares. This means that each individual develops goals, beliefs and feelings of trust and mistrust, but that these feelings are also influenced by his or her membership of a group, culture, online communities, network or social position. While groups and individuals are distinct in trust relationships, they are also eminently interconnected in the form of interaction networks. As we understand it, these questions have been amplified since the emergence of the digital era, where the connections among citizens are more diffuse, digital and multifaceted (depending a lot on an internet connection, and the possibility to exchange and to meet online). The emergence of a digital era has also amplified the questions of disinformation, distrust and conspiracy theories.

V. Trust, Power, and Exclusion

Trust and conflicts are also closely linked with the questions of inclusion and exclusion as exclusion, inequalities, and injustice can lead to conflicts (see Figure 5, Chapter 4).

Hardin and Farell also demonstrate that the question of trust is linked to power and can be affected by extreme power disparities. Power asymmetries would not enable trust relations but instead make them impossible. (Farell, 2004: 89).

In every society, there is discrimination between those who can vote and those who cannot, those who can be judges and those who cannot be.

Inequalities between countries also give them a different weight in the international order. There is therefore a negotiating space in which the norms of justice are produced, a negotiating space in which certain individuals are accepted and others excluded. While those who participate in the elaboration of justice norms will tend to show confidence, those who are excluded will, conversely, develop mistrust.

According to Charles Tilly, the creation of citizenship is inseparable from a process of exclusion between those who benefit from it and those who cannot access it (Tilly 2005: 178)

Referring to the French Revolution of 1789, he demonstrates that, from the moment when only property owners could gain access to citizenship, it clearly began to be based on a relationship of social exclusion. He also argues that, in modern democracies, exclusion is becoming increasingly based on ethnic criteria. This situation would not necessarily give rise to relations of hatred between groups, but it would raise questions of understanding, cohabitation, cooperation and, consequently, trust between groups, entities or communities. Further studies would be needed to extend this analysis to the digital world of today.

In their quest for identity, individuals and social groups alike draw distinctions between "us" and "them," between "I" and "he," by assigning categories to different groups of individuals, distinguished not only by physical attributes—male or female, tall or short, black or white—but also by social attributes—rich or poor, etc. (Spencer et al. 1998: 1140). There are often antagonisms between the different groups, between "us" and "them," with "us" commonly associated with positive values and "them" with negative ones. Stereotyping thus takes the form of inverted mirrors (Jahoda et al. 2007: 193). For several centuries, Africa and its inhabitants were depicted by explorers and observers of the time, including the philosopher Hegel, as "an upside-down world, that is, a world where everything is repulsively different from Europe" (Jahoda et al. 2007). Africa was portrayed as a continent of slaughter and cannibalism, where the union between the sexes was intended to produce as many children as possible for sale. Similarly, an observer of

Japanese culture in the sixteenth century noted over four hundred differences between this culture and Western culture. More recently, the polarizing narratives against migrants both in Europe and in the United States can be analyzed through this lens. During the American 2024 electoral campaign, Donald Trump's speeches indeed fueled rumors suggesting that Haitian communities in Springfield were eating dogs and cats.[10]

The driving force behind these semantic antagonisms is the desire to express difference. (Hodge 1988) This desire would stem from the needs of specific groups to create internal solidarity and exclude others as anti-groups, holders of anti-languages, anti-thought, anti-cultures and anti-worlds. In the case of cultures from other continents, antagonisms were obviously used to establish the superiority and domination of white Western cultures. Stereotypes thus have a structuring function on the surrounding world in the psychology of individuals (cognitive aspect), but they also have repercussions on the behavior and actions of groups (collective aspect). In the case of narratives against migrant communities, a desire for violence and hatred is also not to be excluded as an incitement to hatred. In the case of the Donald Trump Campaign in the United States, the rumors against the Haitian communities in Springfield (Ohio) have indeed led to violence.

Divisions within the same society are not intangible but rather based on representations, and they evolve, but they nevertheless have a decisive influence on trust. Governments and institutions appear to act as arbiters and to form a balance between the different groups, but whether intentionally or not, it asserts the cultural, political or administrative supremacy of certain groups over others. In a lot of countries, ethnic or minority negation is akin to favoring an urban, cultured elite in the capital city, close to the government, to the detriment of other cultures and regions. The "forgotten culture" for instance the middle class in the United States, may be tempted by populism and tend to believe that isolationism could restore their status and dignity (Goodhart 2017, Hansen 2019, Wilkinson 2019, Rodríguez-Pose 2018, Oswald 2022).

Halbwachs's sociology is part of the post-Rationalist Positivism generation of intellectuals who, like Durkheim and Kant, believed in progress.

---

[10]   <https://www.newyorker.com/news/q-and-a/the-historical-precedents-to-trumps-attacks-on-haitian-immigrants>

For Maurice Halbwachs, working-class people feel they have no part to play in collective life and are constantly excluded from it. They are therefore distrustful of institutions and power.

At the other extreme, the role of the bourgeoisie, in a relationship of trust, is to adapt social life to the aims of society (Marcel, 2004–53) At the time of social crises, agents find themselves in positions for which their socialization does not predispose them or which do not correspond to their expectations, which he describes as "positional misery (Bourdieu 1993)." Because it is not possible to negotiate in a climate of mistrust, the only way out is through conflict or revolt. The recent movements of the yellow vests in France can be analyzed through this lens. What the sociologists have not analyzed yet is how new technologies have become a new tool for power. The balance of power will shift toward the countries and groups holding enough data and programming the algorithms. Poor and middle class in this new system may feel lost or exploited and have a feeling of senselessness.

Major social changes may emerge from the current crisis. One of the "refuge" for people faced with these uncertainties and changes could be religion. For Ted Gurr, religion is equally an important feature in the identification of ethnicity if it is one of the characteristics that makes it possible to define a group in its own right, according to the criteria of its members, or according to the criteria of outsiders (Gurr 1993:3).

All post-Cold War religious movements contrast the emotional warmth of intercommunity relations with the abstraction of human relations in societies dominated by bureaucratic organization, and even more so in the disassociated, unstructured societies, which appear to be even more disconnected since the emergence of social media. In most cases, these movements concern populations who are now cultural minorities in secularized societies where, as in some Arab countries, the threat of secularization is perceived as the destruction of identity (Etienne 2002:31).

According to some authors, religious movements primarily serve as political substitutes (Marty et al., 1991: 4). In reality, this resurgence of religion coincides with the collapse of communist ideology and the struggles for decolonization. It is also linked to the end of economic growth and the unfulfilled promises that came with it. It is as if the crisis of the future is bringing back an idealized past.

As Jonathan Fox points out, religion and the coexistence of religions do not in themselves generate conflicts or crisis but can contribute to conflict when certain conditions are met, notably the use of religion to mobilize groups around motives that are not religious in nature, and the role of religious elites in mobilizing certain groups (Fow 1999).

However, the new emerging movements articulated around the claims from the "Global South" for instance during the latest COP29 show that imbalance in power and injustice also play a role in international negotiations over climate with rising tensions, and the idea that money transfers only would solve the problems linked to poverty, lack of infrastructure and climate change, as we have seen in the Chapter 4. Unfortunately, these claims, are not constructive in the way that they prevent rich countries from reducing their climate-related Greenhouse Gas (GHG) emissions, and because they are also powerless to change the international order. They on the contrary strengthen the rich countries attachments to the current systems and institutions as a token of their power. This also illustrates the decreasing trust in institutions that is also found in democratic societies.

VI. The Decreasing Trust in Institutions

The future of democracies as systems and models is at stake in many countries. Will illiberal democracies become more prevalent in the near future? While misinformation, polarization, and social media manipulation largely contribute to these challenges, we must also examine the social factors driving citizen apathy. Several explanations emerge: the increasing complexity of our world and citizens' lack of tools to understand it, social isolation through digital devices, decline of the middle class, rising inflation, economic stagnation, and the erosion of values across political, institutional, and business spheres.

While technological and political factors are critical, social forces play an equally significant role in declining democratic engagement. A major challenge is modern life's growing complexity. As global issues become more interconnected, many citizens lack the tools and knowledge to understand them fully. This overwhelm can lead to disengagement, as people feel powerless to influence outcomes, particularly with globalization.

Social isolation, worsened by widespread digital device use, compounds this problem. Though technology connects us virtually, it weakens face-to-face interactions and community bonds. Strong democracies need active citizens who participate in civil society, yet the digital age has eroded these vital relationships.

Economic conditions fuel citizen apathy. Middle-class decline, rising inflation, and economic stagnation have created widespread dissatisfaction and insecurity. When people struggle with basic needs, they're less likely to engage in democracy. The erosion of core values, accountability, transparency, and integrity—in political, institutional, and business spheres further distances people from the system.

Economic inequality and stagnation, and the subsequent reduction of trust, drive democratic decline. The widening gap between wealthy and ordinary citizens breeds resentment and suggests a system favoring elites. Rising costs of living alongside stagnant wages intensify these frustrations.

The shrinking middle class poses a particular threat. Historically, this group has stabilized democracies by supporting fair and inclusive policies. As it diminishes, so does its moderating influence, creating opportunities for populist or authoritarian leadership.

The erosion of fundamental values across political, institutional, and business sectors poses a crucial yet understated threat to democracy. When accountability, transparency, and integrity become optional rather than essential, public trust crumbles. Citizens increasingly view leaders and institutions as self-serving rather than working for the common good.

Poor ethical leadership in both public and private sectors undermines systemic trust. When leaders prioritize quick wins over lasting stability, they foster a spiral of distrust that can weaken democratic foundations.

Conclusion

In his farewell address from the White House on January 15, 2025, President Joe Biden expressed deep concerns about the state of American democracy, stating:

> *Americans are being buried under an avalanche of misinformation and disinformation, enabling the abuse of power. The free press is crumbling. Editors are disappearing. Social media is giving up on fact-checking. The truth is smothered by lies told for power and*

profit. It erodes a sense of unity and common purpose. It causes distrust and divisions. Participating in our democracy becomes exhausting and even disillusioning.

This served as a strong warning.

Trust is the bedrock of institutions, economies, and political systems. In democracies, trust is essential, but it must remain dynamic—allowing for media scrutiny, public criticism, and political opposition. Unlike authoritarian regimes, democracies cannot rely on propaganda and deception to manufacture trust; instead, they must cultivate it through transparency, accountability, and institutional integrity.[11]

Meanwhile, many autocratic regimes consolidate their power and advance their disinformation campaigns in the form of foreign interference. Across the world, pro-democracy movements have faced brutal repression, while within established democracies, disinformation campaigns and extremist actors actively work to erode public confidence in democratic institutions. This makes democracies uniquely vulnerable to both internal subversion and external interference. The challenge of restoring trust in democracy is particularly acute in an era of instantaneous information, where perception can be shaped as rapidly as reality.

Compounding this challenge are geopolitical tensions and proxy conflicts, where even democratic nations sometimes resort to actions—such as unauthorized strikes on foreign soil—that raise serious ethical and legal concerns. Such actions, while often justified as necessary security measures, risk undermining international law and democratic credibility.

Yet despite these pressures, democracies remain strongholds of fundamental freedoms, particularly in Europe, where institutions continue to protect rights such as freedom of expression, speech, and movement. However, recent shifts in asylum and migration policies threaten to erode long-standing human rights protections. Moreover, the rise of illiberal and populists including far-right movements in countries like Germany, Belgium, France, Italy, Romania, Bulgaria, and Austria signals an ongoing struggle over democratic values.

---

[11] The very nature of trust in democratic systems fundamentally differs from that in autocratic rule.

Democratic resilience is undeniable. Democratic systems have repeatedly demonstrated their ability to withstand attempts at destabilization. However, they remain inherently fragile, as their legitimacy ultimately depends on maintaining sufficient public trust. Without it, the very foundations of democracy are at risk. The Donald Trump administration's efforts to sow distrust in federal institutions—through disinformation, exaggerations, and sarcasm—risk seriously undermining American democracy and the other democracies worldwide.

But imagine a world where free elections gave way to abuse of power and the end of freedoms, while citizens remained apathetic.

References

Almond G. and Verba S. *The Civic Culture*, Princeton editions Princeton University Press, 1963.

Barber B. *The Logic and Limits of Trust*, New Brunswick, Rutgers University Press, 1983.

Bauer M., Cahlíková J., Chytilová J., Roland G., Želinský T. "Shifting Punishment onto Minorities: Experimental Evidence of Scapegoating," *The Economic Journal*, Volume 133, Issue 652, May 2023, 1626–1640, <https://doi.org/10.1093/ej/uead005>

Bernoux P. and Servet J. M. *La construction sociale de la confiance*, Paris, éditions Montchrestien, 1997.

Bianco W. T. *Trust, Representatives and constituents*, Michigan, Michigan Press of University, 1994.

Bollmann, H. S., & Gibeon, G. *The spread of hacked materials on Twitter: A threat to democracy? A case study of the 2017 Macron Leaks* (Doctoral dissertation, Hertie School), 2022.

Bourdieu P. *La misère du monde*, Paris, Le Seuil, 1993.

Braithwaite V., Levi M. *Trust and Governance*. Russell Sage Foundation Series on Trust, New York, 1998.

Burch S., Shaw A. Dale A., Robinson J. "Triggering transformative change: a development path approach to climate change response in communities" in *Climate Policy*, 2014, Vol. 4, 467:487 <https://doi.org/10.1080/14693062.2014.876342>

Cappella J. N., Jamieson K. H. *Spiral of Cynism*, New York University Oxford Press, 1997.

Cervi, L., Carrillo-Andrade, A. "Post-truth and disinformation: Using discourse analysis to understand the creation of emotional and rival narratives" in *Humanitas: revista científica de comunicación*, 10(2), 2019, 125–149.

Dasgupta P. "Trust as a commodity" in Gambetta D. *Trust: Making and breaking cooperative relations*, Oxford, Blackwell publishers, 1988, pp. 49–72.

Dasgupta P. "Trust as a commodity" in Gambetta D. *Trust: Making and breaking cooperative relations*, Oxford, Blackwell publishers, 1988, pp. 49–72.

De Araujo Arosa Monteiro R., Ferraz de Toledo R., Roberti Jacobi R. "Dialogue Method: A Proposal to Foster Intra and Inter-community Dialogic Engagement." *Journal of Dialogue Studies, Special issue, Dialogue with and among the Existing, Transforming and Emerging Communities*, 2021 Vol. 9, 165–183, <https://doi.org/10.55207/FWCB1722>

Deutsch M. "Trust and Suspicion," *Conflict Resolution*, Number 2 (Vol. 8) 1958.

Dogan M., Highley, J. *Elites, crises and the origins of regimes*, New York, Rowman and Littlefield, 1998.

Dorenspleet R. "The structural context of recent transitions to democracy," in *The European journal of political research*, May 2004, N° 43, Vol. 3.

Dunn J., *Trust and political agency*, United States Harvard University Press, 2000.

Etienne B. *Islam, Les Questions Qui Fâchent*; Hachette, 2002.

European Parliament 2022. "Hungary can no longer be considered as a democracy" MEPs: Hungary can no longer be considered a full democracy | News | European Parliament (europa.eu) https://www.europarl.europa.eu/news/en/press-room/20220909IPR40137/meps-hungary-can-no-longer-be-considered-a-full-democracy

Fine G. A. "Rumor, Trust and Civil Society: Collective Memory and Cultures of Judgment", *Diogenes* 2007, 54 (1):5–18. <https://doi.org/10.1177/0392192107073432>

Foley M and Edwards B. :"Is it time to disinvest in social capital?." in *Journal of Public Policy* n° 19, 1999, pp. 141–174.

Fukuyama F: *Trust: The Social Virtue and the creation of Prosperity.* New York, Free Press, 1995.

Funtowicz S. "From risk calculations to narratives of danger," in *Climate Risk Management*, 27, 100212, 2020. <https://reader.elsevier. com/reader/sd/pii/S2212096320300024?token=B6CBEA02054999EE 756B04F666D4701C3905CDF460510AD1F52DF26A57DFF692EC2 12672A58C4D7A49A536076F984FAB&originRegion=eu-west-1&origincreation=20210808171814>. Last accessed 08.08.2021.

Gaborit P. *Restaurer la confiance après un conflit civil*, L'Harmattan, 2009 a.

Gaborit P. "La confiance après un conflit ou la confiance désenchantée," in Bertho A., Gaumont-Prat H. et Serry H. *Colloque international La confiance et le conflit*, Université Paris Vincennes Saint Denis, 2009b.

Gaborit P. *Learning from Resilience Strategies in Tanzania, an Outlook of International Development Challenges*, Peter Lang International, 2021, <https://doi.org/10.3726/b18824>

Gaborit P. "Resilience and Climate Disaster Management in Cities: Transformative Change and Conflicts", *Journal of Peacebuilding & Development*, special issue, Nov. 2022, (2022a) <https://doi. org/10.1177/15423166221128793>

Gaborit P. "Climate Adaptation to Multi-Hazard Climate-Related Risks in Ten Indonesian Cities: Ambitions and Challenges" in *Climate Disaster Risk*, 2022 Vol. 37, 100453, (2022b) <https://doi.org/10.1016/j. crm.2022>.100453

Gaborit P. (Ed). "*Climate Adaptation and Resilience: Challenges and Potential solutions. Anticipatory governance, Planning and Dialogue*", Peter Lang, 2022c.

Gamson W. A. *Power and Discontent*, Belmont Dorsey, 1968.

Giddens A. *The consequences of modernity*, Stanford, Stanford University Press, 1990.

Giddens A.: *Modernity and self-identity*, Stanford, Stanford University Press, 1991.

Girard R. *The Scapegoat*, Johns Hopkins University Press, 1986.

Goodhart D. *"The road to somewhere: The populist revolt and the future of politics"* London, Hurst and Company, 2017.

Grabner-Kräuter S. "Empirical Research in Online Trust. A Review and Critical Assessment", *International Journal of Human-Computer Study.*2003 <https://doi.org/10.1016/S1071-5819(03)00043-0>

Hamm, J. A., van der Werff, L., Osuna, A. I., Blomqvist, K., Blount-Hill, K. L., Gillespie, N., … Tomlinson, E. C.: Capturing the conversation of trust research. Journal of Trust Research, *14*(1), 1–7. 2024 <https://doi.org/10.1080/21515581.2024.2331285>

Hardin R.: "The street-level-epistemology of trust", *Politics and Society n°21*, 1993, pp. 505–529.

Hardin R. *Trustworthiness* New York, Russel Sage foundation, 1996.

Hardin R. (Ed): *Trust and Trusworthiness.* New York, Russel Sage foundation editions, Series on trust, volume 4, 2002.

Hardin R. (Ed): *Distrust*, NYC, Russell Sage Foundation, 2004.

Harris L. M. Chu E., Ziervogel G.: "Negotiated Resilience." EDGES, Institute for Resources, Environment and Sustainability, University of British Colombia, 2017.

Hersh M. A. "Barriers to ethical behaviour and stability: Stereotyping and scapegoating as pretexts for avoiding responsibility," *Annual Reviews in Control*, Volume 37, Issue 2, 2013, 365–381, <https://doi.org/10.1016/j.arcontrol.2013.09.013>

Hetherington M. J. "The political relevance of political Trust", *American Political Science Review*, n°92, 1998: 791–808.

Hodge R. and Kress G. *Social Semiotics.* Cambridge, Polity, 1988.

Inglehart R. *Modernization and Post modernization: cultural, economic and political change in 41 societies*, Princeton, Princeton University Press, 1997.

Khodyakov D. : "Trust as a process: A Three-dimensional Approach", *Sociology*, London, 2007, vol. 41 pp. 115–132.

King K., Wang b. "Diffusion of real versus misinformation during a crisis event: A big data driven approach", *International Journal of Information Management.* 71. 2023 <https://doi.org/10.1016/j.ijinfomgt.2021.102390>

Koivula, A., Malinen, S., & Saarinen, A. "The voice of distrust? The relationship between political trust, online political participation and voting", *Journal of Trust Research*, 11(1), 59–74. 2021. <https://doi.org/10.1080/21515581.2022.2026781>

Lazuech G. *"Toute confiance est d'une certaine manière confiance aveugle,"* Pleins Feux, Variations n°9 2002 <https://www.librairie-sciencespo.fr/livre/9782847290004-toute-confiance-est-d-une-certaine-maniere-confiance-aveugle-gilles-lazuech/>

Levi M. "Social and unsocial capital", *Politics and Society n°24*, 1996 pp. 45–55.

Lijphart A. The *politics of accommodation: pluralism and democracy in the Netherlands*, Berkeley, University of California Press, 1968.

Lijphart A. *Democracy in plural societies*, Yale éditions New Haven: university press, 1977.

Lind M. *The new Class War: Saving Democracy from the Managerial Elites*, New York, Portfolio Penguin, 2019.

Livet P. *Gouvernance et confiance'* lecture series, Institut des nouvelles technologies de Namur, January 20, 2007.

Luhmann, N. *Trust and Power: Two Works by Niklas Luhmann*. 1979, Translation of German originals Vertrauen 1968 and Macht 1975, Chichester: John Wiley, 1979.

Marcel J. C. "Les derniers soubresauts du rationaliste durkheimien : une théorie de l'instinct social de survie chez Maurice Halbwachs," in Deloye Y. et Haroche C. *Maurice Halbwachs: Espaces, mémoires et psychologie collective*, Paris, édition publications de la Sorbonne, Sciences Politiques, 2004.

Misztal B. A. "The Notion of Trust. Social Theory," *Policy, Organisation and Society*, 5(1): 6–15. 1992. <https://doi.org/10.1080/10349952.1992.11876774>

Möllering G. *Trust, Reason, Routine, Reflexivity*, Oxford, Elsevier, 2006.

Moravcsik A. "Taking preferences seriously: A Liberal Theory of International politics," *International Organization*, vol. 4, n°51, fall 1997, p. 513–533.

Nordlinger E., *Conflict regulation in divided societies*, Cambridge editions Cambridge University Press, 1972.

Nyhan R. C. "Changing the Paradigm Trust and Its Role in Public Sector Organizations." *The American Review of Public Administration* Vol. 30, March 2000, pp. 67–105.

Offe C., *Modernity and the state East West*; Cambridge, Cambridge Press, MIT, 1996.

Orléan A. "La théorie économique de la confiance et ses limites," in Laufer R. and Orillard M (dir), *La confiance en questions*, Paris, édition Harmattan, collection Logiques Sociales, 2000, pp. 59–79.

Oswald M. *The Palgrave Handbook of Populism*, Palgrave, 2022. <https://link.springer.com/book/10.1007/978-3-030-80803-7>

Putnam R., Leonardi R and Nanetti R. *Making democracy work: civic tradition in modern Italy*, Princeton University Press, Princeton, 1993.

Putnam R. D. "The strange disappearance of civic America," *American Prospect*,n° 24,1996, pp. 34–49.

Rawls J. *Political Liberalism*, New York, University Columbia Press, 1996.

Rose R. "Russia as an hourglass society. A constitution without citizens", *East European Constitutional review* 1995, pp. 34–42.

RSF Reporters Without Borders, World Press Freedom Index 2024.

RSF Reporters Without Borders, World Press Freedom Index 2025.

Ruscio K. P. "Jay's Pirouette, or Why Political Trust is not the Same as Personal Trust", *Administration and Society*, November 5, 1999, n°31 Vol. 5 pp. 639–657.

Sa'adah A. "Regime Change: Lessons from Germany on justice, Institution building and democracy," in *Journal of conflict resolution*, N° 50, volume 3, June 2006.

Scott K. D. "The causal relationship between Trust and the assessed value of management by objectives", *Journal of management*, vol. 6, 1980, pp. 157–175.

Seligman A. *The problem of Trust*, Princeton, Princeton University Press, 1997.

Simmel G. *The Sociology*. New York Free press, 1964.

Six F. E., Latusek D. "Distrust: A critical review exploring a universal distrust sequence," *Journal of Trust Research*, 13:1, 1–23, 2024 <https://doi.org/10.1080/21515581.2023.2184376>

Stedman S. J. "Spoiler problems in Peace Processes." *International Security*, Vol. 22, n°2, Autumn 1997.

Stedman S. J. "Peace Processes and the challenges of violence," in Darby J. and Mac Ginty R., *Contemporary Peace Making: Conflict Violence and Peace Processes*, London and New York, Palgrave-Mac Millan editions, 2003.

Szescynski B. "Risk and Trust: The Performative Dimension", *Environmental Values*, The White Horse Press, Cambridge UK, 1999, pp. 239–252.

Sztompka P. *Trust a sociological theory*, New York, Cambridge University Press, 1999.

Sztompka P. "Trust and Distrust and Two Paradoxes of Democracy", *European Journal of Social Theory*, 1(1) 19–32.

Thuderoz C., Mangematin V. et Harrisson D. *La confiance: approches économiques et sociologiques*, Paris, édition Gaëtan Morin, 1999.

Thuderoz, C. *Négociations: essai de sociologie du lien social* Paris, PUF, 2003.

V-Dem Institute, 2023, *"The democracy report"* 2023.

V-Dem Institute, 2024, *"The democracy Report"* 2024.

Warren M. E., "Deliberative democracy and authority", *Revue de Sciences Politiques américaine* 1990, pp. 46–60.

Wheeler N. Trust Building in international relations, Peace Prints 2011 <https://wiscomp.org/peaceprints/4-2/4.2.9.pdf>

Zuiderveen Borgesius, F. "Discrimination, artificial intelligence, and algorithmic decision-making." Council of Europe, Directorate General of Democracy, 2018, 42.

CHAPTER 2

New Technologies and the Use of Power

"The telescreen received and transmitted simultaneously. Any sound that Winston made, above the level of a very low whisper, would be picked up by it; moreover, so long as he remained within the field of vision which the metal plaque commanded, he could be seen as well as heard. There was of course no way of knowing whether you were being watched at any given moment. How often, or on what system, the Thought Police plugged in on any individual wire was guesswork. It was even conceivable that they watched everybody all the time. But at any rate, they could plug in your wire whenever they wanted to. You had to live—did live, from habit that became instinct—in the assumption that every sound you made was overheard, and, except in darkness, every movement scrutinized."

—Georges Orwell, 1984

The excerpt above about the telescreen is from the first chapter of *1984*. It is part of the initial descriptions of Winston Smith's daily life in Airstrip One, where Orwell introduces the oppressive tools of the Party, including the telescreen. This device, which monitors both visual and audio activity, is symbolic of the Party's omnipresent surveillance and the lack of personal privacy in Orwell's dystopian world. This passage highlights how new technologies, when controlled by centralized power, can be weaponized to monitor and manipulate individuals. The telescreen—a fictional but prescient device— serves as a metaphor for invasive surveillance technologies that strip people of privacy and autonomy. Orwell's depiction warns how unchecked technological advances can become instruments of oppression instead of progress.

* * * * * *

History has been shaped by battles, revolts, revolutions, and developments. The twentieth century was marked by two world wars, the opposition of two blocs during the Cold War, and the race toward nuclear and thermonuclear weapons. These events defined an era of intense conflict and rapid technological advancements aimed at gaining military and ideological superiority.

The twenty-first century, although more peaceful, is characterized by uncertainty, the rise of multipolar powers, and the spread of new technologies capable of controlling citizens and minds. Never in history has it been possible to control so many populations so effectively and comprehensively. Instruments such as the internet, created to enhance freedom and facilitate communication, have been transformed into tools of power by governments, businesses and malicious actors.

The question then arises: What are the main characteristics of these technological developments, and how have they been used? Unfortunately, research has not yet unveiled the full impacts of the use of new technologies by democratic and non-democratic regimes. However, there have been some attempts to define the main trends and characteristics of such uses. Surveillance technologies, data analytics, AI, and social media manipulation are among the key tools employed by some regimes. Countries like China or Iran have exemplified the control of citizens down to their private lives and intimacies, employing a vast network of surveillance cameras and advanced facial recognition technologies to monitor and suppress dissent (Strittmatter 2021). But the same technologies are also used in Western countries, and in developing countries for instance in Rwanda.

The social credit system is another example of technology's potential to control and influence behavior. This system assigns scores to citizens based on their behavior, rewarding compliance and punishing deviance. Is this the distorting mirror of a world shaped by technology? Are we on the road to voluntary servitude, willed and organized by digital giants and an omnipresent security ideology? The above examples are not the exotic scarecrow of another world, but the vision of what technology allows when implemented without control.

Digital surveillance and control extend beyond dictatorships and autocratic regimes. Many countries have embraced various forms of technological monitoring. Russia has invested heavily in tracking online activities and suppressing opposition voices. Middle Eastern governments have similarly used technology to restrict dissent and control information flow. Even democratic nations face mounting concerns about the security-privacy balance as governments and corporations collect vast amounts of data on citizens.

European governments have developed protective regulations, though these may prove inadequate as technology advances. The EU's General Data Protection Regulation (GDPR) represents one effort to protect personal data and privacy. Yet the rapid evolution of technology demands updated regulations to address emerging challenges. The international community needs to collaborate on establishing norms and standards that both prevent technology misuse and foster its benefits. The EU AI Act marks Europe's initial step in this direction.

The benefits of emerging technologies are indeed often emphasized to show how they can enhance citizens' lives, such as by improving cost-effectiveness in services, advancing healthcare, and more (Bostrom, 2014: 77–78; Schwalbe & Wahl, 2020). However, new technologies also bring risks and may cause harm to individuals and society. Identifying these potential harms is challenging because they are based on statistical probabilities, which are difficult to determine given the limited empirical data available on a technology's potential toxicity (Elliott et al., 2011: 3). Even when researchers try to distinguish between statistical risks and mere uncertainties surrounding the technology, their interpretations of the data remain subjective and open to debate (Elliott et al., 2011: 10).

Assessing the risks of technologies like AI is crucial for ensuring their optimal implementation, with some experts advocating for delaying AI adoption until harms and risks are thoroughly understood (Grans, 2024). This chapter will identify the main trends including cyber threats, the Internet of Things (IoT), the 5G and 6G and the increased vulnerabilities in part I, go through AI and appliance of AI with other technologies such as surveillance and face recognition that have sparked significant concern due to their potential impact on citizens leading to some experts have called "surveillance capitalism" (Zuboff 2020). It will also address the technologies including high risks (part II) including the storage of biometric data, the facial recognition and the massive surveillance and their applications in warfare (e.g., in Ukraine) and their potential use by governments in peace times.

This chapter will demonstrate that, while new technologies hold the promise of improving our lives in countless ways, they also carry the risk of being used for control and oppression. It is imperative that we remain vigilant and proactive in ensuring that these powerful tools are used responsibly and ethically. The lessons from autocratic regimes serve as a stark reminder of what can happen when technology is wielded without checks and balances. The future of technology and freedom depends on our collective efforts to strike the right balance between innovation and regulation.

I. Main Trends

The dream of a just and free cyberspace was short-lived. In 1996, John Perry Barlow (1947–2018) issued a declaration of independence for the internet: "We must call for a civilization of the mind in cyberspace, more humane and just than that of governments" (Gomart, 2020). However, as new technologies emerged, so did the global imperative to collect and control data. Experts now consider globalization and digitization intrinsically linked. Between 2005 and 2016, the flow of online data multiplied by 80.

Major digital platforms have permeated every sector of the economy. Their terms and conditions now regulate our daily lives alongside national legislation. Platforms like Google, Facebook, and Amazon have become so deeply woven into our routines that their policies have a large impact on societies.

For governments, data collection represents security and sovereignty issues, and for businesses, it is essential to value creation. For users, they offer new services at the risk of privacy or anonymity. The balance between the benefits of these services and the potential risks to individual freedoms and privacy is a delicate one. This dynamic has led to a continuous debate over the rights of individuals versus the needs of states and corporations.

This omnipotence is accentuated by the rise of AI. States are becoming aware of what has been termed "surveillance capitalism" (Gomart, 2020: 161), which is based on the extraction and use of personal data without individuals' knowledge using methodology like online "scrapping" (Zuboff, 2020). This form of capitalism relies heavily on the ability to predict and influence human behavior, creating a new kind of economic and social power structure.

Individuals can now be profiled instantly. We have moved from mass consumption to personalized consumption, creating instantaneous reflexes. This increasingly leads to intimate consumer profiling (Gomart, 2020:164, Tonon 2020, Smuha 2023), made possible by the perfect control of continuous flows of personal information captured and quantified by algorithms. These algorithms can predict our preferences, behaviors, and even our future actions with astonishing accuracy, raising ethical and moral concerns.

European public authorities find themselves in the role of regulator in the data war being waged by China and the United States, with the global Data Protection Regulation (GDPR), the Digital Services Act, and the AI Act. Using a metaphor, digital platforms are taking on traditional government functions, while states increasingly operate as networks. This shift has led to an interweaving of resources and created decentralized systems—both public and private—that constantly gather personal data (Gomart, 2020).

Such interconnection fundamentally transforms power relationships between public institutions and citizens, as well as between countries.

Data and Privacy

Protecting data is crucial because it helps safeguard personal privacy, ensures financial security, and maintains trust in digital systems. When data is compromised, it can lead to identity theft, financial loss, and severe breaches of privacy. On a larger scale, data protection is essential for maintaining national security and protecting critical infrastructures. If sensitive information is leaked or misused, it can disrupt services and even pose threats to national safety. Trust in digital systems and online transactions also hinges on data security. In essence, robust data protection measures are fundamental for individual privacy, financial security, national safety, and overall trust in digital platforms.

At the same time, organizations can increasingly expect to be affected by data leakage from compromised AI models, AI-powered malware, and phishing. Data mining has become a new instrument for power, with data being considered as "the new oil" and being collected by large companies as well as by public authorities.

In 2000, the EU and the USA signed the Safe Harbor agreement, which authorized the transfer of European citizens' data to American companies. This agreement was invalidated by the Court of Justice in 2015 (Schrems I ruling). The Privacy Shield, its successor, was invalidated by the same court (Schrems II ruling) in July 2020, highlighting the lack of recourse for European citizens when their data is transferred and exploited in the United States, notably by the FBI Federal Bureau of Investigation and the NSA National Security Agency. It took 15 years for Europeans to grasp the scale of the problem. American groups have benefited from a vast, less-regulated domestic market. The more fragmented EU has not been able to benefit from the same effect of scale nor develop high-performance services. It reacted in 2018 with the GDPR, which aims to "give citizens back control of their personal data." But belatedly adopted, and poorly applied, the GDPR fails to protect sensitive data. The problem is both regulatory and industrial, as Europe lacks world-class cloud hosts to protect its data, such as health and education data (Gomart, 2020: 165, Tonon 2020, Wach et al. 2023).

The GDPR represents an attempt to provide a regulatory and legal response to the EU's political and industrial needs. However, its enforcement and effectiveness have been inconsistent across member states, creating varying levels of data protection and compliance. This inconsistency undermines the GDPR's core objective: creating a harmonized data protection framework across the EU (Solove 2022, Smuha 2023). Furthermore, the rapid pace of technological advancement often outpaces regulatory frameworks, creating gaps and loopholes that can be exploited.

Cyberattacks, Hybrid Attacks, and Foreign Interference

Cyberattacks are computer-to-computer attacks that undermine the confidentiality, integrity, and availability of computers and their stored information (Kim et al. 2012). These attacks are increasing and have become commonplace. For example, on October 7, 2024, a pro-Russian group launched a cyberattack on Belgian cities and provinces, making their systems unavailable for half a day.[12] The attackers' goal was not to steal (personal) data, but to disrupt systems. Earlier this year, Belgian federal government departments were the victims of such an attack that lasted almost a day. The attack could be traced back to Russian by the Belgian Centre on Cybersecurity. These collectives have been targeting several European countries supporting Ukraine in recent years, and this specific attack was considered as a response to Belgium's promise to buy Caesar guns and donate them to Ukraine.

In 2024, Europe faced a series of cyberattacks from hackers in Russia, China, and other countries, targeting politicians, parliaments, and other institutions.[13] Europe's awareness of such threats began with the 2007 cyberattack on Estonia, followed by the widespread NotPetya attack and "WannaCry" ransomware in 2017.

Securing cyberspace has become a critical challenge due to the sophistication and severe impact of these attacks. Modern attacks involve complex coordination among hackers across international borders. Their motivation has evolved from thrill-seeking to financial and military gains, creating a

---

[12] <https://www.vrt.be/vrtnws/en/2024/10/07/pro-russian-group-launches-cyber-attack-on-belgian-cities-and-pr/>

[13] <https://www.politico.eu/article/europe-cyberattacks-russia-china-uk-ministry-of-defence-hacks/>

self-reinforcing cycle (Kim et al. 2012). Making matters worse, most cyber-criminals evade punishment, as their technical expertise often surpasses that of the international authorities tasked with stopping them.[14]

Cyberattacks are launched by coordinated groups of criminals from different countries, each with specific expertise. For example, one person writes spam emails, another controls a botnet (a network of compromised computers used for attacks), another manages servers that collect stolen data like names and credit card numbers, while another converts this information into money and distributes it among group members.[15]

According to security agency experts, investigating these cases often requires coordination between law enforcement agencies from multiple countries. While the EU remains fragmented, recent cyberattacks have shown that cybersecurity agencies from different countries can restore compromised systems.

Cyber threats, however, pose significant risks to national security and critical infrastructure, potentially exposing sensitive policy documents, strategic plans, and weapons systems to adversaries. The disruption of French railways ahead of the 2024 Summer Olympics demonstrates this vulnerability.

These threats emerge from various sources—foreign governments, transnational criminal organizations, and terrorist groups.

AI is also making cyberattacks and cybercrime more sophisticated, posing significant risks to citizens. A cyberattack refers to any effort to alter, disrupt, or destroy computer systems, networks, or the information they contain (Waxman, 2011: 422). AI is fueling the rise of a new generation of cybercriminals who can launch highly targeted and complex attacks at an unprecedented speed and scale. These AI-powered attacks can bypass traditional rule-based security systems, utilizing what is known as "offensive AI" to remain undetected (Guembe et al., 2022:. 2).

AI can also be applied in numerous ways to enhance cyberattacks. Two prominent methods include the use of self-learning malware and automated domain generation, which allow attacks to evolve and adapt in real time.

---

[14] <https://www.scworld.com/feature/border-crossing-fighting-international-cyber-crime>
[15] <https://www.scworld.com/feature/border-crossing-fighting-international-cyber-crime>

AI is also being leveraged for intelligent profile targeting, enhanced intelligence gathering, malware optimization. (Guembe et al., 2022, p. 19–21).

Of particular concern for everyday citizens is the use of AI-driven password cracking, which directly threatens personal data security. AI can be employed in three primary methods of password attacks: brute-force attacks, password guessing, and password theft (Guembe et al., 2022, p. 22). One alarming example is the application of an AI system called assGAN. In a real-world scenario, assGAN successfully cracked 24.2 percent (10,478,322) of 43,354,871 unique passwords from the LinkedIn data breach using only a common password dataset known as RockYou—without any prior access to LinkedIn's datasets (Guembe et al., 2022, p. 22). This demonstrates the devastating potential of AI-driven attacks to compromise personal security on a massive scale without the person being attacked even noticing that it is happening.

Generative AI can equally help criminals and scammers copy official documents, create fake voices and even make fake videos based on real images found on the internet. For instance, brouteurs, or online scammers, exploit vulnerable individuals through fake profiles and deceptive narratives. With advancements in technology, particularly AI, these scams have become more sophisticated, using realistic AI-generated images and automated conversations to build credibility and manipulate victims. The risks include financial losses, emotional trauma, identity theft, and even reputational damage, as scammers may misuse personal information.

European experts have identified three main trends (Gehem et al. 2015).

· The emergence of a new cybercrime economy (e.g., with ransomware and phishing)
· The reliance of the state economies on large internet platforms
· Hybrid warfare is becoming prevalent with cyberattacks, but also the use of cyber weapons

Cyber-powered weapons have seen a significant increase during the 2024 Israel/Lebanon conflict, notably Israel's strategic use of compromised paging devices that were planted nearby Hezbollah leaders before being detonated. These sophisticated cyber-enabled tools represent a new frontier in modern warfare technology. Cyber weapons now serve multiple critical functions, primarily in enemy force localization and advanced facial recognition systems. While such technological warfare tools have been deployed in the Ukraine conflict, they have been used extensively in military operations involving Israel, Hamas, and Hezbollah, marking a significant evolution in contemporary warfare tactics. However, despite their theoretical potential to save lives, cyber weapons have not prevented massive human rights violations in Gaza and Lebanon and the occurrence of war crimes.

As a conclusion, cybersecurity has become an increasing challenge, used for hybrid warfare with cyberattacks and disruptions. It involves foreign governments, malicious actors, but also "cyber mercenaries" and makes the prosecution of actors extremely complicated, which explains that the authors of cyberattacks remain invisible and get unpunished. The U. S. and European Countries have created strong cybersecurity agencies and are acting reactively to restore the connections when they are faced with attacks. Cyberattacks and cyber security are therefore becoming one of the real challenges in the use of power and in geopolitics. Both the attacks and the security systems are likely to infringe privacy and individual rights. And the media are of course not always entirely informed of all processes, most of which remain invisible. That being said, since 2022 and the invasion of Ukraine, European National governments like Latvia, Germany, France and Belgium have tried to communicate better on cybersecurity and cyber threats. It is expected that the 5G, the IoT, and the use of AI will strengthen the hybrid threats and the cyberattacks. Cyber Ark (2024:7) predicts an unparalleled increase in the volume and sophistication of attacks, as skilled and unskilled bad actors can leverage AI to intensify their assaults. The types of attacks will be harder to detect as AI will automate and personalize the attack process. At the same time, the IoT, 5G and 6G, could enhance the impacts of the cyberattacks.

The Internet of Things (IoT)—5G and 6G

By 2030, an estimated 500 billion objects could be connected through the IoT. A key challenge lies in capturing and leveraging these new streams of high-value and industrial data, with national groups receiving priority. Computing power plays a crucial role in this transformation. 5G technology delivers signals directly to users at speeds ten times faster than previous generations, enabling the management of smart cities, connected cameras, industrial and transport automation, autonomous vehicles (Gomart, 2020:166), and telehealth applications. The technology's reliance on trusted suppliers has sparked debates in Western countries about the reliability of Huawei, the leading Chinese provider.

The European Court of Auditors categorizes the providers of 5G in risks categories and highlights the vulnerabilities created by the 5G systems both in terms of sovereignty (dependence of national governments and countries on the providers) and in terms of data protection. The providers from China are categorized as "high risks" but the European Court of Auditors also highlights the difficulties of several European countries in implementing the 5G system, and therefore acknowledges that there are still important efforts to be made in terms of system's resilience and protection (European Court of Auditors 2022)

The implications of 5G extend beyond faster internet speeds. This technology is expected to revolutionize various sectors, including healthcare, transportation, and manufacturing. Smart cities, powered by 5G, will be able to monitor and manage resources more efficiently, improving the quality of life for their inhabitants. However, the increased connectivity also raises concerns about security, privacy, and the potential for misuse of data. The potential for 5G to enable a vast array of new applications and services also underscores the importance of ensuring that the infrastructure and networks supporting this technology are secure and resilient while being deployed.

The deployment of 5G technology also has indeed significant geopolitical implications. Countries are vying for leadership in 5G technology and infrastructure, recognizing its strategic importance. The competition between the United States and China over 5G dominance is particularly notable. The US has expressed concerns about the security risks posed by Chinese companies like Huawei, which has led to restrictions and bans on the use of Huawei equipment in several countries. This geopolitical rivalry over 5G technology is not just about economic competition but also about national security and technological sovereignty. The emergence of the 6G is reinforcing this trend.

6G, or sixth generation wireless technology, is the next step in the evolution of mobile communications, following 5G. While still in the early stages of development, 6G aims to provide even faster speeds, lower latency, and greater capacity than its predecessor. With the ability to handle more data simultaneously, 6G will support a larger number of connected devices and more demanding applications. 6G will enable new applications such as advanced virtual reality, augmented reality, and the Internet of Everything (IoE), where everything is interconnected and intelligent.

6G is expected to revolutionize various industries, including healthcare, transportation, manufacturing, and entertainment, by providing seamless and ultra-fast connectivity. It will also add vulnerabilities to the existing vulnerabilities and enhance the possibilities for malicious actors and for foreign interference, reinforcing the need for Europe (and for other countries) to create an independent internet protected cloud system (e.g., The European Cloud Initiative) and a separate data infrastructure.

II. Artificial Intelligence

AI is emerging as a critical tool of power in the twenty-first century (Gomart, 2020:169). Both nations and private entities are developing strategies—some aiming to gain a competitive edge, others striving to maintain dominance. AI offers unprecedented analytical capabilities, enabling faster and more efficient decision-making across diverse sectors. A key distinction exists between generalist AI and specialized AI: the former includes generative AI and large language models, while the latter encompasses applications like facial recognition, natural language processing, and autonomous vehicles. AI has been likened to the new electricity, capable of applying algorithms to virtually any type of digital data and executing tasks exponentially faster than humans. However, these capabilities come at a significant environmental cost, with AI systems requiring vast amounts of energy for training and operating large-scale models.

Machine learning, a core component of AI, involves training systems on vast datasets to identify patterns and make predictions. This technology is accelerating robotization and the automation of industrial economies, while investment grows in innovations like digital twins and advanced design software. Large online platforms are reshaping industrial dynamics, positioning manufacturers as subcontractors that produce goods on-demand, ordered in real time (Baldwin, 2019). AI is also driving a transformation in the workforce, creating a new division of labor with implications for employment and skill requirements.

Finally, AI development relies heavily on key materials and components, particularly semiconductors. The United States continues to dominate semiconductor production, creating a strategic dependency and intensifying technological competition with China.[16] Despite these advancements, the energy consumption and environmental impact of AI technologies remain critical challenges, underscoring the need for sustainable approaches as AI continues to shape the global landscape.

In this sense, DeepSeek is an interesting example. DeepSeek is a Chinese AI chatbot which has surged in popularity, becoming the most downloaded free app in the United States shortly after its release in January 2025 just being the Chinese New Year and right after the launch of the Stargate AI program in the United States. DeepSeek R1 model, which is now open source, claims to rival OpenAI's advanced O1 model. The development team of DeepSeek says they have achieved this with a budget under $6 million, a stark contrast to the multibillion-dollar investments in AI by U.S. tech giants like OpenAI. The breakthrough lies in its development using reduced-capacity Nvidia H800 chips, sidestepping reliance on high-end Nvidia A100 chips that have been constrained by U.S. export bans.[17]

Whereas OpenAI was founded 10 years ago, has 4,500 employees, and has raised $6.6 billion in capital. DeepSeek was founded less than 2 years ago, would have 200 employees, and would have been developed for less than $10 million raising questions on possible copy of technology.

This accomplishment reflects broader implications in the global AI race, especially as the United States has implemented strict export controls targeting China's access to advanced semiconductor. Restrictions include export bans on chips and chipmaking equipment since October 2023, and a first-of-its-kind prohibition on U.S. persons aiding Chinese chip development

---

[16] In 2018, China's trade deficit in semiconductors exceeded that of hydrocarbons and represented 12 percent of its imports. 75 percent of AI chips are produced by the American company Nvidia. TSMC (Taiwan Semiconductor Manufacturing Company) is at the heart of the Sino-American technological confrontation. The Chinese authorities have launched a plan to catch up by 2025, while the United States seeks to maintain its advantage. The European company ASML located in Veldhoven in the Netherlands, is also dependent on American technology.

[17] – Adam Kobeissi (27.01.025) From: <https://x.com/TKL_Adam/status/188365720 0851362018>

– <https://www.deepseek.com/>

– <https://www.scientificamerican.com/article/why-deepseeks-ai-model-just-became-the-top-rated-app-in-the-u-s/>

without a license. These measures aim to limit China's advancements in both AI and military modernization, and for the United States to maintain technological supremacy.[18] However, DeepSeek's breakthroughs demonstrate that resourceful methods can counter such constraints, potentially diminishing the effectiveness of U.S. policies. While this raises questions about America's long-term strategic position, it also emphasizes the potential for less resource-intensive, climate-friendly AI innovations. This could be an opportunity for other companies to imitate this success, and this could be especially true as the developed system is shaped with open access.[19] DeepSeek-R1 is indeed an open-source AI model. The company has released it under the MIT license, ensuring clear open access for the community to utilize and build upon its model weights and outputs. This open-source approach allows developers worldwide to modify and integrate the model into various applications, promoting innovation and collaboration in the AI community. Eric Schmidt, former CEO of Google, has commented on DeepSeek's significance. In a recent op-ed, he stated that DeepSeek's emergence marks "a turning point" in the global AI race.

DeepSeek's rise isn't without serious ethical and geopolitical concerns. Critics have highlighted the tool's alignment with Chinese state propaganda. Examples include labeling Taiwan's independence movement as a "challenge to China's sovereignty" and avoiding contentious topics like the repression of opponents and Uyghurs, through evasive or malfunctioning responses. Furthermore, data security remains a core concern, as Chinese companies are legally required to provide data to the government upon request. These fears echo the ongoing debate over TikTok and underscore the implications of a major Chinese competitor in AI not just as a technological force but also as a tool of state influence. Because of this, several experts and commentators have already called for President Trump to take action on this matter by banning DeepSeek in the United States.[20]

---

[18] – <https://www.lemonde.fr/economie/article/2025/01/27/la-start-up-chinoise-deep-seek-cree-une-onde-de-choc-sur-le-secteur-de-l-ia_6518928_3234.html?lmd_medium=al&lmd_campaign=envoye-par-appli&lmd_creation=android&lmd_source=-default>

[19] <https://www.nature.com/articles/d41586-025-00229-6>

[20] – <https://www.skynews.com.au/business/tech-and-innovation/chinas-deepseek-ai-is-a-malfunctioning-national-security-threat-filled-with-misinformation-and-propaganda/news-story/219c35e4f6c8596776e5521ae85bcfe7>

The development of AI will fundamentally shape the future of our lives, work, and the global order. It is increasingly evident that superpower rivalry has sparked a high-stakes global race to dominate this transformative field. While recent advancements have marked a win for China in this race, it would be premature to declare that the Chinese are winning the broader AI battle. Free-tier ChatGPT users, impressed by DeepSeek's performance, believe they are witnessing a revolutionary jump in AI functionality. However, this perception is likely results from OpenAI having restricted its comparable high-end models to paying customers for some time.

On the ethical front, growing ties between major tech companies such as Meta and the U.S. federal government are generating concerns among users. Supremacy of U.S. tech giants in the AI race might seem less than ideal. Still, China's undeniable deeper ties between technology companies and the Chinese Communist Party (CCP) creates a data security environment that deepens ethical concerns about censorship and disinformation. This underscores a sobering reality: neither American nor Chinese dominance in AI seems optimal when viewed through the lens of user trust, freedom, and ethical integrity, although for now are American companies favorable.

The AI companies in other countries, including the ones located in European countries, could remain ignored in this race.

The potential applications of AI are vast and varied. In healthcare, AI can assist in diagnosing diseases, personalizing treatments, and managing medical records. In the financial sector, AI can detect fraudulent activities, predict market trends, and automate trading processes. In education, AI can provide personalized learning experiences, track student progress, and optimize administrative tasks. AI's influence extends to other areas, such as environmental monitoring, where it can analyze data from sensors to track pollution levels and predict natural disasters. In agriculture, AI can optimize crop yields by analyzing soil conditions, weather patterns, and pest activity.

However, the rapid development and deployment of AI also pose significant challenges. Ethical considerations, such as bias in algorithms, accountability,

---

<https://www.skynews.com.au/business/tech-and-innovation/chinas-ai-deepseek-calls-for-taiwan-to-be-returned-malfunctions-when-questioned-about-uyghur-muslims/news-story/dbb0f8df9726d73fd98c79a3bfe76d88>

and transparency, need to be addressed. The potential for job displacement due to automation raises concerns about the future of work and the need for reskilling and upskilling the workforce. Additionally, the concentration of AI capabilities in a few countries and companies could exacerbate global inequalities. Policymakers, civil society and digital companies must work together to create frameworks that ensure AI is developed and deployed responsibly, ethically, and inclusively.

AI systems introduce unique risks that are not present in traditional software applications. These risks arise from the complexities of machine learning algorithms and their operational environments (Steimers & Schneider, 2022: 1). Some of the uncertainties worth mentioning when assessing AI as a technology, our chapter will go over potential harm regarding algorithmic biases, privacy and lack of data use consent, and the appliance of AI in human roles can cause a lack of accountability when errors are made.

Algorithmic Biases and Prejudices

The fear of algorithmic bias causing harm is a major concern in the deployment of AI technologies. While AI systems learn from available data, the assumption that this would result in neutral representations of the real world has proven flawed. In reality, algorithms can intensify existing inequalities tied to socioeconomic status, race, ethnicity, religion, gender, disability, or sexual orientation (Panch et al., 2019 : 1–2). For instance, search algorithms have been found to reinforce gender disparities: a search for "CEO" yielded only 11 percent of female results, even though women comprised 20 percent of U.S. CEOs at the time. This highlights how, despite the overrepresentation of men in such roles, algorithms may further amplify these disparities (Panch et al., 2019: 2). This issue, known as "algorithmic repression" or "algorithmic injustice" in the literature, raises concerns that machine learning (ML) systems may perpetuate or exacerbate racial and gender biases (Howard, 2010). A compelling case example from Fu et al. (2020) illustrates how ML algorithms could be used in loan approval processes can lead to discriminatory decisions. These algorithms, focused solely on assessing repayment risk, may disregard fairness considerations, resulting in biased outcomes. The concern is that algorithmic biases can produce real-world harm, further deepening societal divisions along racial, gender, and other identity-based lines. But this is not the only concern.

Privacy Concerns

The collection of geolocation data by apps has sparked numerous scandals, raising serious privacy and ethical concerns. Many apps collect precise location data without users' clear consent or understanding, then sell it to third-party advertisers, data brokers, or government entities. Investigations have shown that fitness, weather, and social media apps track individuals' movements, revealing sensitive patterns about their daily lives—including home and work addresses. This data enables invasive advertising, surveillance, and other harmful uses. The problem grows worse due to unclear data collection practices and weak regulatory oversight. Recent high-profile scandals have exposed major risks, from authoritarian regimes misusing location data to accidental exposure through data breaches. These incidents show we urgently need stronger privacy protection, better user control, and greater accountability from app developers and data brokers. AI plays a critical role in analyzing and using this data. AI algorithms process vast amounts of location data to identify patterns, predict behaviors, and deliver precisely targeted advertising or services. For instance, AI can determine a user's habits, preferences, and even social relationships by analyzing visited places, travel routes, and time spent in specific areas.

Location data is often combined with other information—like social media activity, purchase history, and demographics—to train AI models, improving their predictive and decision-making abilities. This merging of data creates powerful tools for companies and governments but raises serious privacy concerns. Without proper safeguards, AI analysis of location data can enable unethical practices like intrusive surveillance, discriminatory profiling, and manipulation of consumer behavior. The combination of AI and location tracking underscores the urgent need for strong data privacy rules and ethical AI development.

Individuals' data privacy indeed faces uncertainty with the emergence of AI, as ML technologies are dependent on large amount of data for their training. Because of this dependent on data, it could potentially be that AI applications process sensitive private data of normal citizens, such as their messages, photos or health records, available on the internet to ensure accuracy of their own product, and without the individuals being aware that their data is being scrapped (Steimers & Schneider, 2022:11). A literature review on

AI research on the health field found that very few of the papers described how the researchers used and accessed large data sets such as electronical health records in an ethical manner or addressed issues of informed consent (Schwalbe & Wahl, 2020: 1583). Increased precautions and awareness of the topic is needed when introducing AI to different fields of society to ensure our safety and privacy.

Intellectual Property

The rise of AI presents significant challenges for intellectual property (IP) law, particularly in areas such as copyright, patents, and ownership rights. One of the key concerns is the authorship and ownership of AI-generated content—can an AI system be credited as an inventor or creator, or do rights belong to the developers, users, or the entity that owns the AI? This legal gray area has led to disputes over AI-generated art, music, literature, and even inventions. Additionally, AI's ability to train on vast datasets, often scraping copyrighted material without explicit permission, raises concerns about fair use, plagiarism, and the compensation of original creators. In industries like publishing, film, and software development, AI-generated content threatens to dilute originality and undermine traditional copyright protections. As AI continues to evolve, policymakers face the difficult task of balancing innovation and creativity with the protection of IP rights, ensuring that AI development does not come at the expense of human creators and legal integrity.

Transparency and Accountability Concerns

The lack of transparency when it comes to the inner workings of AI-applications has been a concern when applying AI. The "black box" nature of algorithms refers to a situation where the internal workings or decision-making processes of an algorithm are opaque or not easily understood by users, regulators or even the developers themselves (Panch et al., 2019: 2). The potential misuse of AI for example, manipulation of information, scraping personal information from social media, or other actions that could repress or unduly influence the public in a democratic society underscores importance of transparency of the inner functions of the AI algorithm, because without sufficient transparency of the AI, it usage faces limitations of accountability of misuse (European Parliament DG for External Policies of the Union, 2021: 13).

Identity Breaches

The problem of identity breaches facilitated by AI is multifaceted and increasingly concerning as AI technologies evolve. Here are some key issues. AI can be used to create highly sophisticated phishing attacks. By analyzing vast amounts of data, AI can craft personalized and convincing phishing emails that are more likely to deceive recipients. AI tools can automate the process of scraping personal information from social media and other online sources. This data can then be used to build detailed profiles of individuals, making it easier for cybercriminals to impersonate them. AI can generate realistic deepfake videos and audio recordings, which can be used to impersonate individuals convincingly. This technology can also create synthetic identities that combine real and fake information, complicating the detection of fraudulent activities (IMI 2024).

AI can also enhance the sophistication of cyberattacks by identifying vulnerabilities more efficiently and automating complex attack strategies. This makes it easier for attackers to breach systems and steal identities (CyberArk 2024).

Finally, AI tools can automate the process of scraping personal information including pictures from social media and other online sources. This data can then be used to build detailed profiles of individuals, making it easier for cybercriminals to impersonate them.

AI Applied to Surveillance

AI can also be intentionally misused, particularly in surveillance technologies, where governments may employ AI to restrict citizens' freedoms through increased monitoring (Kaplan, 2020: 153).

In Europe, concerns about privacy in biometric data storage were sharply highlighted by the case of facial recognition technology (FRT). Some companies made vast facial recognition databases that were openly available to anyone willing to pay, with no restrictions on how the data could be used (Ahmed, 2023: 66). The core issue with biometric data collection lies in the fact that facial data acts like a digital fingerprint, uniquely identifying individuals. Each face scan they collect strengthens the accuracy of the algorithm, making it easier to identify someone—making the technology very intrusive (Ahmed, 2023: 67).

According to the company's own reports (Clearview AI, n.d.), their AI's biometric database for instance, now would hold over 50 billion images. This extensive data collection has led to significant legal repercussions, such as the 30.5 million euro fine imposed by the Dutch Data Protection Authority (Dutch DPA) for multiple violations of the GDPR, particularly the company's lack of transparency (Autoriteit Persoonsgegevens, 03.09.2024). The Dutch DPA criticized the company for failing to adequately inform individuals that their images and biometric data were being collected and used. Moreover, individuals whose data is stored in the database are entitled to know what information the company holds about them. Despite this, the Dutch DPA found that Clearview AI had been uncooperative in responding to requests for data access, further deepening concerns about the company's disregard for privacy rights (Autoriteit Persoonsgegevens, 03.09.2024). This case exemplifies the growing tension between the rapid development of biometric technologies and the fundamental rights to privacy and informed consent, and the concern for methods of storage of biometric data.

The AI ACT. The European Regulation of AI

Finally, the AI Act (EC, 2021, b) is the first-ever legal framework on AI, which addresses the risks of AI and positions Europe to play a more visible role globally, although it is considered as incomplete.

The AI Act aims to provide AI developers and deployers with clear requirements and obligations regarding specific uses of AI. The AI Act is part of a wider package of policy measures to support the development of trustworthy AI, while possible misuses have been identified by experts in the recent years (Smuha 2021). The AI Act ensures that Europeans can trust what AI has to offer. The risks of AI have indeed been identified as critical by most of the governments, but the EU is the first actor to regulate on the matter. Interestingly, the EU AI ACT proposes a model based on a pyramid, identifying the level of risks (Witzel 2021).

Limited risk refers to the risks associated with lack of transparency in AI usage. The AI Act introduces specific transparency obligations to ensure that humans are informed of the use of AI when they are targeted, therefore fostering trust. For instance, when using AI systems such as chatbots, humans should be made aware that they are interacting with a machine so they can

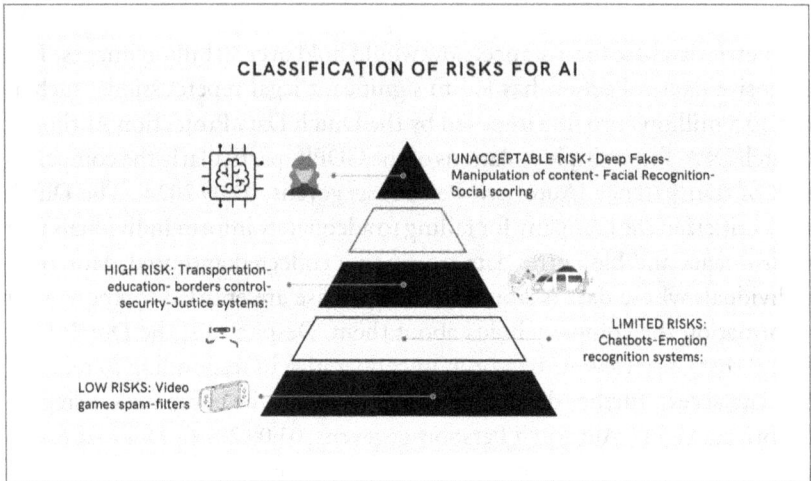

CLASSIFICATION OF RISKS FOR AI

UNACCEPTABLE RISK- Deep Fakes- Manipulation of content- Facial Recognition- Social scoring

HIGH RISK: Transportation- education- borders control- security-Justice systems:

LIMITED RISKS: Chatbots-Emotion recognition systems:

LOW RISKS: Video games spam-filters

Figure 2. Classification of risks for AI- Source author-inspired by Witzel 2021

make an informed decision to continue or step back. Providers will also have to ensure that AI-generated content is identifiable. Indeed, AI-generated text published with the purpose of informing the public on matters of public interest must be labeled as artificially generated. This also applies to audio and video content constituting deep fakes.

Systems identified as high-risk include AI technology used in: critical infrastructures (e.g., transport), that could put the life and health of citizens at risk; educational or vocational training, that may determine the access to education and professional course of someone's life (e.g., scoring of exams), safety components of products (e.g., AI application in robot-assisted surgery); employment, management of workers and access to self-employment (e.g., CV-sorting software for recruitment procedures); essential private and public services (e.g., credit scoring denying citizens opportunity to obtain a loan); law enforcement that may interfere with people's fundamental rights (e.g., evaluation of the reliability of evidence); migration, asylum and border control management (e.g., automated examination of visa applications); administration of justice and democratic processes (e.g., AI solutions to search for court rulings). As mentioned earlier, the display of generated AI images needs to be tagged with the mention "AI" on any information published online, which should normally prevent the use and spread of "untagged" deep fakes.

Most importantly, some uses with unacceptable risks will be banned: for instance, with the prohibition of real-time biometric identification by law enforcement authorities in publicly accessible spaces, but with some notable and clearly defined exceptions.

The text has been finally approved by the EU institutions in March 2024. Critics of the EU AI Act argue that its stringent regulatory framework could hinder innovation and competitiveness within the European economy. While the Act aims to ensure ethical and trustworthy AI, its broad scope and complex compliance requirements may disproportionately burden small and medium-sized enterprises (SMEs) and startups, which lack the resources to navigate these regulations. Some fear that these rules could slow down the adoption of AI technologies across industries, limiting Europe's ability to compete with tech giants from the United States and China, where regulatory approaches are often more flexible. Additionally, the classification of certain AI applications as "high-risk" could discourage investment in key sectors like healthcare, transportation, and finance, where AI has transformative potential. Critics also point out that the lack of global harmonization in AI regulations may fragment markets and create trade barriers, further isolating the European economy in the race for technological leadership. While the Act seeks to strike a balance between innovation and ethical safeguards, many worry it could stifle growth and push AI development outside of Europe.

It is also important to note that the disinformation or malevolent intentions are not really part of the risk assessment, which rather tries to create a framework on future usages and automations. This AI Act is very much expected to provide a safe framework for the reasonable use of AI. It is, however, expected that loopholes will be used, and that it cannot prevent malevolent uses. For instance, Facial Technology Recognition is prohibited un public spaces, with some exceptions and it is already used by companies. Some additional challenges related to the high-risk technologies are highlighted below.

III. High Risks Technologies

Numerous technologies available today can be used by national governments or private groups in ways that undermine fundamental freedoms and human rights. A key example is Facial Recognition Technology FRT. While the EU AI Act bans its use in public spaces, it remains legal in many countries, including the United States, despite some legal challenges. In Europe, companies can still deploy FRT under specific restrictions. It is also used in the war in Ukraine to identify enemies (Mysyshyn 2024) or in China for security and surveillance (Strittmatter 2021).

Looking ahead, a crucial question is whether, in an increasingly interconnected world, the widespread use of these technologies by global powers will inevitably spill over, challenging the EU's ability to enforce its AI regulations and protect its citizens from facial recognition surveillance. Additionally, the growing use of mass surveillance, including FRT, by autocratic regimes raises concerns about its impact on a significant portion of the world's population.

Facial Recognition Technology (FRT)

FRT refers to biometric systems that automatically detect and record a person's face, making it possible to identify or otherwise recognize a person from these digital images. (Mysyshyn 2024) There are several ways in which this technology can be used to identify individuals. One is to use biometric data, for instance measuring the distance between different points on the face, the shape of the eyes, and the width of the nose. Another method is to analyze facial characteristics (Mysyshyn 2024). Facial recognition can be combined with video surveillance, and goes beyond, as it can continuously scan, identify and track individuals. Moreover, the collection of biometric data through FRT, introduces a new level of concerns on privacy and use of personal data, including taken from online images, without the consent of individuals.

The biometric data that FRT gathers and stores, not only can be used to identify a person's identity based on their physical and behavioral characteristics but is also more permanent and personal than any other information. They can be used to uniquely identify a person in different situations, and over extended periods of time, as well as to determine characteristics such as their age, gender and ethnicity (Mysyshyn 2024). Even more worryingly is the fact that individuals have no control over how their biometric data is used, stored and shared, which raises significant ethical and legal concerns, particularly in terms of consent and data protection.

There are foremost responses and regulations:

China has widely adopted FRT, often for public security purposes. The country has implemented regulations to ensure data security, but these are generally less stringent on privacy compared to Western standards. The Personal Information Protection Law (PIPL), enacted in 2021, includes provisions for biometric data but allows extensive use by the government. (Wang et al. 2021)

The EU has been proactive in regulating facial recognition and biometric data. The GDPR sets strict guidelines on data protection and privacy, including biometric data. The AI Act also limits the use of biometric data and prohibits facial recognition in public spaces except in certain circumstances. However, private companies already make use of technology to recognize their employees for computer access for instance. Quid of supermarkets and of administrations in the future?

In the United Kingdom, the court of appeal ruled that facial biometrics are "intrinsically private" and comparable to fingerprints and DNA and should be considered as sensitive information.

In the United States, the regulation is more fragmented. Some states, like California and Illinois, have specific laws governing the use of biometric data. For instance, the Illinois Biometric Information Privacy Act (BIPA) requires companies to obtain explicit consent before collecting biometric data. It also provides legal recourse for mishandling, setting a precedent that other states start to follow. There is however no federal regulation yet (Wang et al. 2024).

In Australia: The Privacy Act 1988 includes provisions for biometric data, and the government has guidelines for the use of FRT. Finally, in Canada, the Personal Information Protection and Electronic Documents Act (PIPEDA) governs the use of biometric data in the private sector, requiring consent and transparency.

Military Use of Facial Recognition Technology

As highlighted by several experts., the FRT has been used in Ukraine since March 2022, through the use of the American company Clearview AI to gather evidence of war crime, to identify Russian war perpetrators, to identify movements at the border, and at checkpoints (Dave et al. 2022). The collection of data relied on satellite imagery, some of which was purchased from private companies, as well as on videos and photos on social media platforms such as TikTok (Mysyshyn et al. 2024). The company Clearview AI has granted Ukraine free access to its software which identifies people by images that have been previously extracted from online social media such as Facebook, Twitter and Vkontakte and search engines such as Google. To identify a person, their photo must be uploaded to the company's biometric database and the algorithm will find a match. The database contains more than 30 billion images and Clearview AI often sells this data to different authorities, mainly law enforcement agencies (Mc Dade 2023).

Existing Risks in Civil Use

The analysis of what is used during military times highlights the dangers and existence of massive databases that are collected through images and videos on the internet. As noted by experts, "The unchecked power that FRT generally offers can be a powerful tool that violate peoples' rights and threatens democracy" (Mysyshyn et al.2024, Cataleta 2024). While some companies such as FindClone, PimEyes, NEC or Search4faces may adhere more strictly to data protection standards, than others, the general risk of abuse and potential for privacy violations is present across all platforms. The use of Clearview AI FRT has sparked debate and concerns, particularly regarding privacy rights, accuracy, and fairness. The reliability of this organization has been questioned by experts and through several court cases (Privacy International 2022). Many experts recognize AI FRT as being an extremely intrusive technology. The organization Privacy International has said that its use represents "a significant expansion of the scope of surveillance, with a very real potential for abuse."

The Storage of Biometric Data

The risks associated with the storage of biometric data, for the use of FRT, or for other purposes are massive. They could be used by private lobbies to intimidate activists. Indeed, evidence has been found that the American company VFluence has stored massive data including personal data of people whose positions, research, jobs or opinions could be detrimental for the industry of pesticide. The listing included information on the financial situation of the persons, their accommodation, family information and information settings found online (Foucart et al. 2024). Adding biometric data would enable similar companies working for major corporations to track the health of activists, journalists or researchers.

The storage of biometric data, and the use of FRT can also deter people from participating in democratic processes, expressing dissent, or participating in demonstrations. These methodologies are already used massively by autocratic regimes. But it could well later on threaten the foundations of a democratic society, including freedom of speech and the right to protest FRT could be used to spy on dissidents, on activists, but also to manipulate or to intimidate the population or specific groups. (Mysyshyn 2024). It could indeed be used to target certain specific communities, ethnic groups, religions, or opponents. It is also subject to errors and bias.

These risks are compounded by the fact that it is often the companies providing FRT that control which people can access the technology, making the system closed. (Mysyshyn 2024).

Massive Surveillance at the Use of Power

Many State governments continue to engage in secret mass surveillance and communications interception, systematically collecting, storing, and analyzing data from various forms of communication, including emails, phone and video calls, text messages, and websites visited (UN Doc A/HRC/39/29). While some States justify such indiscriminate surveillance as essential for protecting national security, this practice is fundamentally incompatible with international human rights law. The lack of an individualized necessity and proportionality analysis, which is a cornerstone of lawful surveillance under human rights standards, makes these measures impermissible (see A/HRC/33/29, para. 58).

States often rely on business enterprises to collect and intercept personal data, raising significant concerns about privacy and accountability. For example, some States compel telecommunications and internet service providers to grant them direct access to data streams flowing through their networks. These systems of direct access are particularly troubling, as they are prone to abuse and often circumvent critical procedural safeguards. Additionally, many States demand access to the vast quantities of information collected and stored by these companies. Mandatory data retention laws require telecommunications and internet providers to indiscriminately collect and store all traffic data of subscribers and users, covering all forms of electronic communication. These laws undermine individuals' ability to communicate anonymously, create risks of abuse, and increase the likelihood of disclosure to third parties—including criminals, or business competitors—through hacking or data breaches. Such measures far exceed what can reasonably be considered necessary and proportionate under international human rights standards. [...] It seems paradoxical that Elon Musk, the owner of platform X (formerly Twitter) who claims to be a libertarian, had been participating in the U.S. government under the Trump administration even temporarily.

A detailed example of surveillance has been unveiled by a team of investigative journalists which has identified the misuse of the Pegasus software by national governments for the surveillance of dissidents including in other countries. The Pegasus investigation, also known as the Pegasus Project, was an international investigative journalism initiative that exposed the misuse of the Pegasus spyware developed by the company NSO Group. This spyware was originally marketed for tracking serious crimes and terrorism, but the investigation revealed that it was used by various governments to spy on journalists, opposition members, activists, businesspeople, and civilians. The investigation began in 2020 when a list of over 50,000 phone numbers, believed to belong to individuals identified as "people of interest" by NSO Group's clients, was leaked to Amnesty International and the group of investigative journalists "Forbidden Stories." This information was shared with 17 media organizations, leading to a collaborative investigation involving over 80 journalists. The findings, published in July 2021, highlighted the extensive and intrusive nature of the surveillance, which included accessing data, images, conversations, and even the camera and microphone of targeted smartphones. The revelations sparked global outrage and calls for stricter regulations on the use of such surveillance technologies (Amnesty International 2021). The transnational reach of surveillance technology is deeply concerning, as it enables coordinated attacks against dissidents even beyond their home countries. This means activists may find no true safe haven, as they can be tracked and targeted anywhere in the world.

The Use of Technology in China As a Case Study

The interpenetration between states and platforms is visible in China (Strittmatter 2021) and the Chinese authorities have made AI a priority since 2016. Surveillance technology across the country involves: standardized internet, generalized video surveillance, large-scale facial recognition, digitized social control (Haski, 2020).

Massive Surveillance has been inherent to the "smart cities" concept. In 2016, China had 176 million surveillance cameras, compared with 62 million in the USA (Gomart, 2020: 175). These are the eyes of smart cities, with players like Megvii and SenseTime.

Since 2017, the National Intelligence Law stipulates that Chinese companies must support, assist, and cooperate with intelligence services creating de facto a direct connection with the leading party and the government. The internet in China operates autonomously, protected from the outside world by a firewall, or Great Digital Wall. The state controls all networks and data to guarantee social stability and cultural security by controlling content deemed dangerous to Chinese identity. These protections enable immediate censorship of content and anticipate social demands. They also enable the development of a vast economic market that is unified and captive.

China's digital giants include Alibaba, Tencent, and Baidu. In 2011, Tencent launched WeChat, which integrates messaging, payment, and e-commerce. It has 900 million users, 90 percent of whom are Chinese. (China had just 22 million internet users in 2000). Tencent's financial strength enables it to invest in foreign groups.

Founded in 2000, Baidu is China's leading search engine and the world's second largest after Google. The group is investing heavily in AI for autonomous vehicles. Three factors have played a key role: the country's switch to mobile telephony, digital payment in the 2010s, and the collusion between public authorities and digital players to put the Chinese population on data. Alibaba, Tencent, and Baidu have contributed to the digitization of national control of society in areas such as security surveillance, data mining, and the control of public opinion (Strittmatter, 2021: 219).

Roberts (2018) highlights the Chinese government's advanced information control strategies, which extend beyond simple content censorship. These include flooding digital spaces with distracting content—such as overwhelming volumes of information or hashtag hijacking—tactics that AI can amplify to stifle public discourse and limit citizens' ability to communicate freely online (Roberts, 2018).

China is also at the forefront of biometric surveillance, and FRT are used to monitor and control some groups and minorities like Uigurs. Parts of the population face intrusive monitoring through a vast network of AI-driven surveillance that strips away privacy and individual freedoms (European Parliament DG for External Policies of the Union, 2024: 13–15). The Chinese government collects biometric data from FRT in schools, on public transport, and across other aspects of daily life (European Parliament DG for External Policies of the Union, 2024: 20). This system allows the state to closely monitor citizens, controlling their movements, relationships, and even political beliefs.

The country also implemented an official Social Credit Score (SCS), a system that aggregates social and behavioral data through AI to rank individuals and corporations. Citizens are rewarded or punished based on a variety of criteria, effectively linking AI-driven surveillance to social control (European Parliament DG for External Policies of the Union, 2024: 19). This example demonstrates how AI, can be used not just to monitor, but to actively control and shape societal behavior.

The Chinese model of digital control has shown efficiency in various aspects. For instance, during the COVID-19 pandemic, China's use of digital tools for contact tracing, quarantine enforcement, and public information campaigns was highly effective in controlling the spread of the virus. However, this efficiency comes at the cost of individual freedoms and raises concerns about the potential for abuse of power. Kai Strittmatter shows in his book *"We Have Been Harmonized: Life in China's Surveillance State"* (2021) how the massive surveillance system gets more and more invasive for personal freedoms. He is worrying about the impacts on this invasive surveillance on Chinese society and on individuals, as well as the unhidden possibility to store data including biometric data of the whole world population, using the information that is already available online.

Conclusion

In the OECD countries and beyond, the spread of Information and Communication Technologies (ICTs) has favored the monopoly of digital platforms, weakened competition, degraded economies, and widened inequalities. The concentration of digital power in a few companies has led to concerns about market dominance, privacy, and the ethical use of data. Regulatory bodies around the world are grappling with how to address these challenges and ensure that the benefits of digital technologies are shared more equitably.

The rapid pace of technological change continues to outstrip the ability of regulatory frameworks to keep up. As a result, there is an ongoing struggle to balance innovation with the protection of individual rights and societal values. The future of the digital landscape will depend on how these complex and interrelated issues are addressed by governments, businesses, and civil society.

The integration of digital technologies into every aspect of life has created new opportunities and challenges for large businesses to store data and scale up their business. They also have created open paths for malicious actors to develop massive threats and for certain governments (or groups) to control society to the detriment of privacy and human rights. They also transform geopolitical games with transnational cyber threats and foreign interference, as there are no boundaries in connectivity. Digital technologies have reshaped the game in the use of power, changed the geopolitical landscape and have reshuffled the cards to the detriment of individual freedoms and of vulnerable groups.

As we move forward, it is essential to foster a collaborative approach that includes diverse stakeholders from different sectors and regions. This will help ensure that digital transformation is guided by principles of inclusivity, fairness, and sustainability. Additionally, education and public awareness about digital rights and responsibilities will be critical in empowering individuals to navigate and shape the digital world responsibly.

In conclusion, the evolution of cyberspace and digital technologies has brought about significant changes in how we live, work, and interact. While the potential benefits are immense, so are the risks and challenges. By addressing these issues thoughtfully and proactively, we can harness the potential of technology and limit the harms and risks for individuals and societies.

1984 or rather 2024 George Orwell had warned us ... But *Imagine a world where you couldn't trust whether the voice on the phone was truly your loved one or an imitation, or whether the person speaking in a video was actually saying those words—or if it was just a deepfake.*[21]

References

Ahmed, I. "ACLU v. Clearview Ai," *DePaul Journal of Art, Technology & Intellectual Property Law*, *33*(1), 4, 2023 Available at: <https://via.library. depaul.edu/jatip/vol33/iss1/4>

---

[21]   University of Latvia, AI4DEBUNK Webinar, 21/01/2025

Amnesty International. "Uncovering the Truth: The Digital Surveillance Crisis Wrought by States and the Private Sector." 2021. DOC 10/4491/2021.

Anderson J. Rainie L. *Many Tech Experts Say Digital Disruption Will Hurt Democracy*. Pew Research Center. February 2020.

Autoriteit Persoonsgegevens, "Dutch DPA imposes a fine on Clearview because of illegal data collection for facial recognition." 2024 September 3rd <https://www.autoriteitpersoonsgegevens.nl/en/current/dutch-dpa-imposes-a-fine-on-clearview-because-of-illegal-data-collection-for-facial-recognition>

Baldwin R. *"The Globotics, Upheaval, Globalization, Robotics and the Future of Work"* Oxford, Oxford University Press, 2019, pp. 2–12.

Beraja M., Kao A., Yang D. Y., & Yuchtman N, "Exporting the surveillance state via trade in AI" (No. w31676), *National Bureau of Economic Research*. 2023 <https://doi.org/10.3386/w31676>

BIR Business Information Review The *Dangers of Generative Artificial Intelligence*, 2023, Vol. 40(2), 46–48.

Bostrom, N. *Superintelligence: Paths, dangers, strategies*, Oxford, England: Oxford University Press, 2014 <http://site.ebrary.com/lib/alltitles/docDetail.action?docID=10896241>

Bozic V. *The dangers of artificial intelligence*, 2023 <https://doi.org/10.13140/RG.2.2.22058.80326>

Cataleta M. S. "The Fragility of Human Rights Facing AI Human Artificial Intelligence," Working Paper number 2, 2024 <https://www.academia.edu/43891664/The_Fragility_of_Human_Rights_Facing_AI>

Cyberark. *Identity Security. Threat landscape report* 2024.

Dave P. and Dastin J. "Exclusive: Ukraine has started using Clearview AI's Facial Recognition during war", *Reuters* March 13, 2022. <https://www.reuters.com/technology/exclusive-ukraine-has-started-using-clearview-ais-facial-recognition-during-war-2022-03-13/>

Ekman A. "La smart City chinoise : nouvelle sphère d'influence", Ifri study, Ifri December 2019.

Elliott K. C. & Dickson M. *Distinguishing risk and uncertainty in risk assessments of emerging technologies*, 2011. URI: <https://philsci-archive.pitt.edu/id/eprint/8771>

European Parliament: Directorate-General for External Policies of the Union. *Digital technologies as a means of repression and social control. European Parliament*, 2021. <https://data.europa.eu/doi/10.2861/953192>.

European Court of Auditors: *Special Report 03/2022: 5G roll-out in the EU: delays in deployment of networks with security issues remaining unresolved*, 2022. <https://www.eca.europa.eu/en/publications/SR22_03>

European Parliament: Directorate-General for External Policies of the Union. *Digital technologies as a means of repression and social control*. European Parliament., 2021 <https://data.europa.eu/doi/10.2861/953192>.

European Parliament: Directorate-General for External Policies of the Union & Ünver, A. *Artificial intelligence (AI) and human rights: using AI as a weapon of repression and its impact on human rights: in-depth analysis*, Publications Office of the European Union, 2022. <https://data.europa.eu/doi/10.2861/52162>.

Foucart S, DeBre E., Gibbs M. "Révélations sur le fichage à grande échelle de personnalités gênantes pour l'industrie agrochimique", *Lighthouse Reports. Journal Le Monde*. 30. Sept. 2024.

Fu, R., Huang, Y., & Singh, P. V. "AI and algorithmic bias: Source, detection, mitigation and implications", *Detection, Mitigation and Implications July 26, 2020*.

Gaborit P. "A sociological approach to disinformation and AI: concerns, responses and challenges", *Journal of Political Science and International Relations*, 2024, Vol. 7, n°4, 75–88 <https://doi.org/10.11648/j.jpsir.20240704.11>

Gans, J. S. "How learning about harms impacts the optimal rate of artificial intelligence adoption", *Economic Policy*, eiae053, 2024. <https://doi.org/10.1093/epolic/eiae053>

Gehem M., Usanov A., Frinking E., Rademaker M. "Assessing Cyber Security: A meta analysis of threats, trends, and responses", The Hague Center of Security Studies. 2015.

Gomart T.,*Guerres Invisibles*, Tallandier, Collection Texto Essai, 2020.

Guembe, B., Azeta, A., Misra, S., Osamor, V. C., Fernandez-Sanz, L., & Pospelova, V. "The emerging threat of AI-driven cyber attacks: A review", *Applied Artificial Intelligence*, 36(1), 2037254, 2022. <https://doi.org/10.1080/08839514.2022.2037254>

Hansson, S. O. "From the Casino to the Jungle: Dealing with Uncertainty in Technological Risk Assessment," *Synthese* 168(3), 2009, pp. 423–432.

Haski P., prologue in Kai Strittmatter, *"Dictature 2.0 : Quand la Chine surveille son peuple et demain le monde"*, Tallandier, Collection Texto Essai, 2021.

Heldt, A. "Reading between the lines and the numbers: an analysis of the first NetzDG reports", *Internet Policy Review*, 2019, 8(2), 1–18. <https://doi:10.14763/2019.2.1398>

International Management Institute, *Artificial Intelligence, Threats and Security Issues*. 2021.

Kaplan, A "Artificial intelligence, social media, and fake news:: Is this the end of democracy?" in Akkor Gul, A., Ertürk, Y. D. and Elmer Paul (eds.) Digital Transformation in Media & Society: Istanbul University Press, 2020, pp. 149–161.

Khlaaf, H. "Toward comprehensive risk assessments and assurance of AI-based systems", *Trail of Bits*, 7, 2023.

KIM, Seung Hyun; QIU-HONG WANG; and ULLRICH, Johannes B. "A Comparative Study of Cyberattacks", *Communications—ACM*. 55, (3), 66–73. March, Vol. 55 No. 3, 2012 Pages 66–73. <http://dx.doi.org/10.1145/2093548.2093568>

Kissinger H., Schmidt E., Huttenlocher D., *The Age of AI*, John Murray, 2024.

Kusche, I. "Possible harms of artificial intelligence and the EU AI act: fundamental rights and risk", *Journal of Risk Research*, 2024, 1–14.

Lloyd's Register Foundation. "Word Risk Poll 2021: A digital Poll." Gallup 2021.

Martinsen J. "The uncertainties of Emerging Technologies: AI, Facial Recognition, Social Scoring, and Mass, Surveillance," AI4DEBUNK, 2024.

McDade A. "Clearview AI Scraped 30 billion images from Facebook and gave them to Cops. Business", *Insider*, 2023.

Mysyshyn A. "Advanced Technologies in the War in Ukraine: Risks for Democracy and Human Rights," *German Marshall Fund. GMF Report*. Oct. 2024.

Ng. A "Why Artificial Intelligence is the New Electricity", *The Wall Street Journal*, June 9, 2017.

Obama. B. *Wired*. 2016, <https://www.wired.com/2016/10/president-obama-guest-edits-wired-essay>

Panch, T., Mattie, H., & Atun, R. "Artificial intelligence and algorithmic bias: implications for health systems." *Journal of global health*, 9(2), 2019. 010318. <https://doi.org/10.7189/jogh.09.020318>

Privacy International. "The Clearview Ukraine Partnership: How Surveillance Companies Exploit War." March 18.2022.

Report of the United Nations High Commissioner for Human Rights on the Right to Privacy in the Digital Age, UN Doc A/HRC/39/29, August 3, 2018.

Roberts, M. *Censored: distraction and diversion inside China's Great Firewall.* Princeton University Press, 2018.

Schwalbe, N., & Wahl, B. "Artificial intelligence and the future of global health." *The Lancet*, 395(10236), 2020. 1579–1586. <https://doi.org/10.1016/S0140-6736(20)30226-9>

Smuha, N. A. "Beyond the individual: governing AI's societal harm", *Internet Policy Review* 10.3 2021 <https://doi.org/10.14763/2021.3.1574>

Solove D. "The limitations of privacy Rights", *Teach Privacy*, 2022 <https://teachprivacy.com/the-limitations-of-privacy-rights/>

Steimers A., & Schneider M. "Sources of risk of AI systems", *International Journal of Environmental Research and Public Health*, 19(6), 3641, 2022. <https://doi.org/10.3390/ijerph19063641>

Strittmatter S. *Dictature 2.0 : Quand la Chine surveille son peuple et demain le monde*, Tallandier, Collection Texto Essai, 2021.

Tonon C. *"La GovTech, nouvelle frontière de la souveraineté numérique,"* Ifri Study, Ifri November 2020.

Wach K., Doanh Duong C., Ejdys J., Kazlauskaitė R., Korzynski P., Mazurek G., Paliszkiewicz J., Ziemba E. "The dark side of generative artificial intelligence: A critical analysis of controversies and risks of ChatGPT," *Entrepreneurial Business and Economic Review* 2023. <https://doi.org/10.15678/EBER.2023.110201>

Wang X., Wu Y.C, Zhou M. and Fu H. "Beyond surveillance: privacy, ethics, and regulations in face recognition technology. Front", *Big Data* 7:1337465, 2024. <https://doi.org/10.3389/fdata.2024.1337465>

Waxman M. C. "Cyber-attacks and the use of force", *Yale J. Int'l L.*, 36, 421. 2011 Available at: <https://scholarship.law.columbia.edu/faculty_scholarship/847>

Witzel L. "5 Things You Must Know Now About the Coming EU AI Regulation" <https://medium.com/@loriaustex/5-things-you-must-know-now-about-the-coming-eu-ai-regulation-d2f8b4b2a4a9>. 2021 pp. 128–146.

Zuboff S. *"L'âge du capitalisme de surveillance,"* Paris, Zulma, 2020.

CHAPTER 3

Disinformation As a New Threat for Global Geopolitics[22]

Imagine a world where there are no facts, no truth, no science ... A world of lies and manipulation.

Philipp Semmelweis, a nineteenth-century Hungarian physician, is often hailed as a pioneer of antiseptic practices after he demonstrated that hand-washing with a chlorinated solution drastically reduced maternal mortality in hospitals. Despite the empirical success of his findings, his ideas were widely rejected by the medical community of his time, as they contradicted established norms and challenged prevailing medical authority. This historical episode highlights the dangers of disinformation and resistance to evidence-based knowledge. In the case of Semmelweis, the spread of misinformation, coupled with professional arrogance and institutional inertia, delayed the adoption of life-saving practices. This parallels modern challenges where disinformation—often amplified by digital platforms—undermines public health measures, scientific evidence, and trust in expertise. The Semmelweis story is a stark reminder of how rejecting verified information, whether out of ignorance, pride, or vested interests, can have profound consequences for society.

I. Introduction

International organizations have classified disinformation as one of the main threats to modern lifestyle and democracy for over a decade now. Digital technologies ceaselessly reinvent and profoundly reshape modern lifestyles and business environments. As we have seen in the previous chapter, AI is

---

[22] *This chapter takes stock of the results of the project AI4DEBUNK* <www.AI4DEBUNK.com> *The Discussion Points and more background information can also be found in my recent article* A sociological Approach to Disinformation and AI: concerns, responses and challenges *in Journal of Political Science and International Relations, 2024, Vol. 7, nº4, 75–88* <https://doi.org/10.11648/j.jpsir.20240704.1>

bringing a new disruption to how we access knowledge, create, spread, and understand information by blurring the lines between real and manipulated content. The speed at which the information circulates has also been booming, facilitated by seamless connections and technologies. The number of mobile phones circulating worldwide recently reached 7.21 billion. Currently, around 67 percent of the world population has access to the internet, compared to just 1 percent in 1995.

The next stage of technology—including online collaborative platforms, social media, 4G mobile phones, smart devices, and cloud computing—has generated additional transformations. These include connected devices, instant connections, human-device interactions, new forms of information and images, and the creation of online communities.

These changes are likely to impact on geopolitics and the balance of power, as even decision-makers can be influenced. Another important transformation is that social media represent an important source of news for most of their users (King et al. 2023). Connectivity is boundless: collaborative platforms including social media have enabled a direct link among people.

On the other side of reality, rapidly evolving technologies, including AI, are "increasing opportunities to create realistic AI-generated fake content, but also [...] facilitating the dissemination of disinformation, to a (micro) targeted audience and at scale by malicious stakeholders" (Bontridder and Poullet, 2021). Concerns have been raised about copyrights, biased algorithms, business models using massive data to deceive individuals, and the replacement of jobs by technology in numerous AI sectors. AI technologies also facilitate the creation of video, text, and image content based on false information, making it difficult for individuals and the media to trust information (Newman 2024). Disinformation can take different forms, including fake news and impersonation enabled by deepfakes. It is then relayed by bots and amplifiers through automatic dynamic cross-referencing of networks (Bergmanis-Koräts et al. 2024:3). The aim of this chapter is to provide a first analytical approach to the topic based on the current debates by researchers and media-literature and media articles—and on a first analysis of fake news.

This chapter will consider the question of disinformation and see how it is likely to impact the current geopolitical landscape. It will go through the current processes, threads, and impacts of disinformation(I), present case studies and evidence of interference to illustrate the discussion (III) address social media manipulation(IV), coordinated inauthentic behavior and interference in elections (V) before addressing the remaining challenges (VI). We argue in his chapter that there is an increased awareness in European countries about the impacts of disinformation, but also a discrepancy between the identification of "fake news" and disinformation, the available responses and the lack of understanding of the strategies, threads, and actors of disinformation.

II. Processes and Impacts of Disinformation

Questions about trust and credibility in online activities have persisted for over a decade (Grabner-Kräuter, 2006). The study of disinformation gained momentum in 2016 following the scandal involving Cambridge Analytica's interference in the U.S. elections. This topic has attracted increased attention, particularly in what is often called the "post-truth era"—a term that became popular after the 2016 U.S. presidential election, when "fake news" emerged as a common phrase (Grinberg et al., 2019:374). "Fake news" is understood here as a general term used by the media to encompass both disinformation and misinformation. Identifying the processes, threats, and impacts is crucial for understanding this research.

Disinformation is defined as the deliberate dissemination of false, incorrect, or misleading information to cause harm. It is false, inaccurate, or misleading information shared with the intent to deceive the recipient (Bontridder et al. 2021, Persily et al. 2021, Shahbazi 2024). Misinformation, on the other hand, can be defined as the unintentional spread of false, incorrect, or misleading information. In simpler terms, the most important distinction between information, misinformation, and disinformation lies in the question of truth and intent. While information is true, misinformation and disinformation are untrue, with disinformation being intentionally deceptive (Stahl, 2006).

Disinformation is not a new phenomenon; it has deep historical roots stretching back to ancient times when rulers and leaders would intentionally spread rumors or misleading information to weaken opponents or control public opinion (Fine, 2007:6, Schreirer 2023). Although modern technologies have dramatically increased the speed and scale at which disinformation can spread, the tactics remain strikingly similar—relying on manipulating emotions and sowing confusion to achieve a specific agenda.

Before delving into the various efforts aimed at regulating misinformation and disinformation, it is essential to first identify the processes, threats, actors, narratives, and impacts. The European Commission defines disinformation as "verifiably false or misleading information that is created, presented, and disseminated for economic gain or to intentionally deceive the public, and may cause public harm." Public harm comprises threats to democratic political and policy-making processes as well as public goods such as the protection of EU citizens' health, the environment or security. Disinformation does not include reporting errors, satire and parody, or clearly identified partisan news and commentary. European Commission (2018d: 3–4). By disseminating disinformation online, malicious stakeholders may for instance seek to discredit scientists or leaders, to polarize information or to destabilize democratic institutions. The indicators of disinformation are intentional harm; false or misleading content; presence of bias or manipulative techniques; audience targeting including micro-targeting and psychometric profiling. Disinformation can take different forms, which are not restricted to social media although social media manipulation is an important aspect of it. Apart from the social media manipulation it also encompasses false flag operations, influence operations and the manipulation of social media.

The manipulation of social media in particular, aims to create and amplify false or misleading narratives in messages spread through social media platforms. Among other examples, trolls or even trolls' farms using social bots have been implicated in spreading divisive content on the main social media platforms like Facebook and X (former Twitter) and Telegram, aiming to create instability, sow discord and influence public opinion (Hayduchyk et al. 2024, Sessa et al. 2024, Bergmanis-Koräts 2024). Although this manipulation content is not always traced back to its origin, experts have identified several "campaigns" of manipulation on social media, some of which can be traced back to outside of the EU. The role of the Kremlin for instance in creating disinformation campaigns about the war in Ukraine, has uncovered several Telegram channels in Russian amplifying and spreading misleading messages (Dauksas et al. 2024). The disinformation campaigns have more broadly been targeting to stir anger and reactions on divisive elements such as the arrival of migrants, or on the conflicts in the Middle East.

Although most of the platforms include a form of moderation, this has not prevented the spread of false information, with a higher role for TikTok, X and Telegram because of their policies to restrict moderation to a minimum. However, Meta, Instagram and even traditional media have not been spared by the campaigns. In addition to this, most of the platforms have given up on fact-checking and replaced it by Community Notes since the second election of Donald Trump, creating worries on the future impact of disinformation worldwide.

Many platforms have indeed abandoned traditional fact-checking processes in favor of systems like Community Notes, which rely on user-generated contributions to flag or explain misleading content. While this shift appears to democratize content moderation, it raises serious concerns about disinformation's global impact. These systems assume users will provide accurate, unbiased, and informed input. Yet coordinated campaigns, echo chambers, and bad-faith actors can exploit them, potentially amplifying rather than reducing disinformation. This approach weakens accountability as platforms shift moderation responsibility to users, diminishing incentives for robust fact-checking mechanisms. Consequently, fighting disinformation becomes more challenging, especially in polarized or vulnerable societies where false narratives can severely damage public health, democratic processes, and social cohesion. The shift toward decentralized moderation systems may indicate a concerning retreat from the proactive measures needed to tackle the growing disinformation crisis.

Interestingly, most of the moderation is now automatized, and according to Meta (Facebook, WhatsApp), more than 90 percent of the moderation is done by AI tools, the rest being done by human moderation (Kertysova 2018). The moderation did however not prevent massive disinformation campaigns, including media spoofing, impersonation of celebrities (Bergmanis-Koräts et al. 2024) and deep fakes, notably during the Doppelgänger disinformation campaign which started in 2023 (Sessa et al. 2024).

Malicious actors have engaged in false flag operations, where they pose as individuals or groups from different countries to spread disinformation and conceal their true identity. This tactic aims to exploit existing tensions and manipulate perceptions of geopolitical events. These false flag operations are expected to be empowered with the use of AI. The role of large language models and automated dynamic network cross referencing are already massively used by the actors of disinformation (Bergmanis-Koräts et al. 2024, Haiduchyk et al. 2024). Worryingly, further advances in Machine Learning ML will increasingly enable adversaries to identify individuals' unique characteristics, beliefs, needs, and vulnerabilities. This will allow them to deliver highly personalized content and target those most susceptible to influence with maximum effectiveness (Kertysova 2018). These techniques could also create micro approaches, to target decision-makers or voters. Disinformation has been making use of different narratives to undermine Western institutions and democracies. This has taken the form of fake news but also taken the form of false/forged journal and media covers, use of voices, manipulation of images, and use of artificial intelligence. An example is the forgery of the satiric French Journal "Charlie Hebdo" in February 2024 to mock the Ukrainian army command. This occurred simultaneously with the forging of covers of famous newspapers.[23] This trend has enabled the emergence of fake news detectors which have been set up rather as a "reactive" approach as they cannot "prevent" the disinformation to be spread. Another trend is also the increasing acceptance of the use of "fake news" and disinformation as a "political weapon" among democracies as we unfortunately witness in European and in U.S. elections campaigns in 2024.

Several Countries have employed influence operations to shape perceptions and policies in target countries through a combination of disinformation, propaganda, and covert activities (Cull 2009, FA 2017, Charillon 2018…). These operations often target vulnerable populations and exploit societal divisions to advance their interests. Entities from foreign countries have been suspected of interfering in elections in other countries through disinformation campaigns aimed at undermining confidence in democratic processes, spreading conspiracy theories, and supporting divisive political candidates or causes. But the emergence of AI system has created more threats, as shown by the social media influencing used by the company Cambridge Analytica[24] in the U.S. 2016 elections (Wade, 2018). It is indeed nowadays technically

---

[23] Podcast, Colin Gérard, RFI, March 2024
[24] Cambridge Analytica—an advertising company recruited for the campaign of Donald Trump, amassed large amounts of data, built personality profiles for more than 100 million registered U.S. voters and then, allegedly, used these profiles for targeted advertising.

possible to differentiate between demographic and psychometric profiling techniques to influence political elections (Kertysova 2018). Demographic profiling is informational, segmenting voters based on factors like age, education, employment, and country of residence. On the other hand, psychometric profiling is behavioral, allowing for voter segmentation based on personality traits (Wade 2018). Another related trend is automatic content generation.

It is essential to critically evaluate information sources and be cautious of false or misleading narratives, especially in the context of online information consumption and social media engagement. Additionally, ongoing research and monitoring efforts by governments, think tanks, and civil society organizations appear important but not sufficient for identifying and countering disinformation campaigns effectively.

Several states and non-state actors or even groups or individuals can be threat actors regarding disinformation. A significant type of non-state actor is the so-called advanced persistent threat (APT), a term used to describe malicious, organized, and highly sophisticated cyber campaigns. APT groups are often funded by state governments, providing them with the resources to conduct cyberattacks and other hybrid threats like disinformation (Małecka, 2024 : 55). These groups played a notable role during the Russian military invasion in Ukraine, acting as separate entities from the state despite government funding. Russian disinformation about NATO and the war in Ukraine achieves global reach through these non-state actors. Russian "influence-for-hire" firms, such as the Social Design Agency (SDA), the Institute for Internet Development, and Structura, have received substantial funding from Russia to spread disinformation. In response, the European Union imposed sanctions on SDA and Structura, recognizing these campaigns as threats to the EU and its member states (Antoniuk, 2023).

Democracies and in particular Western democracies are increasingly challenged by narratives that exploit deep-seated fears and prejudices, fracturing societies along political, ethnic, gender, and religious lines (Moravcsik 1997, Butcher & Neidhard, 2020: 5, Bollman 2022) and to amplify scapegoating approaches against vulnerable groups to create fear and anger. The scapegoating approach is indeed recognized as one of the strategies of manipulation to polarize the political debate (Deutsch 1958, Girard 1986, Hersh 2013, Goodhart 2017, Betz et al. 2021, Bauer et al. 2023). From the rise of Euroscepticism, which culminated in Brexit, to the divisive rhetoric surrounding migration, gender, and religion, these narratives have often proven effective in manipulating public opinion and creating sharp societal divides. Identifying these polarizing narratives is crucial because they undermine social cohesion, fuel extremism, and threaten democratic stability. It can be used to undermine trust in institutions and in democratic systems, as it is here understood since our first chapter, that trust is an important pillar of stability for societies (Deutch 1958, Luhmann 1979, Putnam 1993, Seligman 1997, Sztrompa 1999, Hardin 2002&2004, Tilly 2005, Gaborit 2009, Hamm et al. 2024). By recognizing and understanding these narratives, society can better counteract their harmful effects, prevent the spread of disinformation, and promote a more inclusive and unified European community.

When addressing the topic of disinformation, discussions often center on prevention strategies, its impact, and methods for detection. However, an equally critical aspect is understanding the threat actors responsible for disseminating disinformation. Since disinformation involves the deliberate spread of false information, the intent behind these actions is to craft and promote a deceptive narrative. By examining the actors who originated the disinformation, we can gain deeper insights into its mechanisms, and how to protect ourselves from it.

The EU classifies threat actors based on whether they are state actors, non-state actors, or proxies, and further categorizes threats by their attribution as either technical or political. When foreign entities engage in the dissemination of false information, it is referred to as "Foreign Information Manipulation and Interference" (FIMI). FIMI is characterized as a "mostly non-illegal pattern of behavior that threatens or has the potential to negatively impact values, procedures, and political processes." This activity is manipulative, intentional, and coordinated, involving both state and non-state actors, including their proxies operating inside and outside their territories (ENISA (i), 2023: 6). In a report assessing the cyber threat landscape, the EU Agency for Cybersecurity (ENISA) identified several primary motivations driving these actors. These motivations include geopolitical aims, intentions to cause disruption, ethical reasoning, and economic or financial gain (ENISA (ii), 2023: 12). Having established that these actors operate with diverse motivations, we will now delve deeper into specific cases and types of threat actors.

As developed in this paragraph, disinformation can take different forms, processes, and threads, and can be relayed by different actors including states, groups or individuals, whereas the quick advancements in technology are amplifying the opportunities of targeted foreign information manipulation and interferences. The impacts can be manifold, and there is a legitimate amplifying concern among civil citizens in Europe. Examples of interference can also be identified.

III. Examples of Interference

Disinformation campaigns are increasingly used to manipulate public opinion, exploit vulnerabilities, and undermine trust. They embed key disinformation narratives, including those related to COVID-19 vaccines, Ukrainian refugees, NATO troops, EU sanctions. These campaigns involve methods such as fabricated stories, impersonation of legitimate sources, and exploitation of real events.

Pro-Kremlin narratives about COVID-19 vaccines promoted Russian alternatives while spreading conspiracy theories, targeting several EU countries. Similarly, fabricated stories about Ukrainian refugees aimed to incite anti-refugees' sentiment and undermine support for Ukraine. Disinformation targeting NATO troops sought to erode support for their presence in Eastern Europe, while false claims about EU sanctions blamed them for global food shortages, deflecting responsibility from Russia.

These campaigns share common goals: to erode public trust, deepen societal divisions, and destabilize target countries. Effective countermeasures, such as media literacy, fact-checking, and strategic communication, are essential to mitigate their impact.

Narratives About COVID-19 Vaccines in Multiple EU Countries (2020–2021)

Media outlets, especially in Russia, spread several false and misleading narratives about COVID-19 vaccines across multiple EU countries in 2020–2021.[25] They claimed that Western vaccines were unsafe or ineffective, while

---

[25] <https://www.reuters.com/world/china/russia-china-sow-disinformation-undermine-trust-western-vaccines-eu-report-says-2021-04-28/>

promoting Russian vaccines like Sputnik V.[26,27] It included allegations that the EU's vaccine procurement process was corrupted or mismanaged.

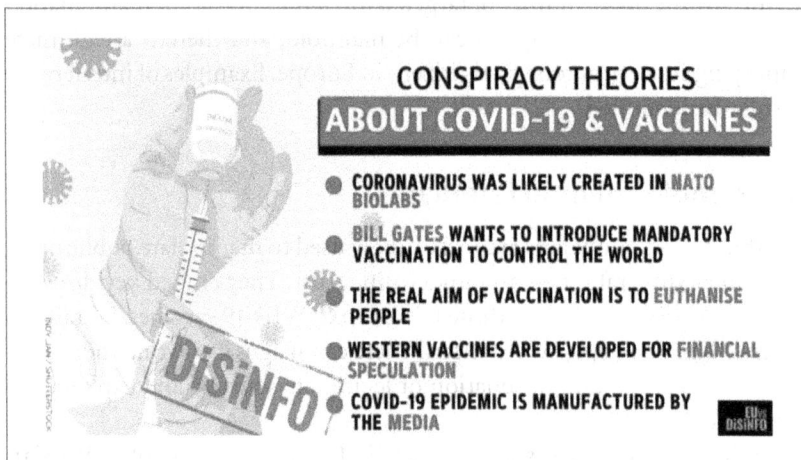

Figure 3. Conspiracy theories about COVID-19 and vaccines. Source: EU DISINFO LAB[28]

Further, conspiracy theories that COVID-19 vaccines were created as biological weapons in secret U.S./NATO laboratories.[29] Common narratives across the countries included fake stories of vaccine injuries, severe harmful effects, and conspiracy theories linking vaccines to totalitarian control, 5G technology, or profiteering by pharmaceutical companies.[30] The narratives that were built exploited citizens' vaccine hesitancy and, thereby, amplifying anti-vaccination sentiments. These disinformation campaigns targeted multiple EU countries, including Germany, France, Poland, and Italy, where millions of social media users were exposed to anti-EU and anti-vaccine propaganda. Serbia, where Russian state media promoted positive coverage

---

[26] <https://www.cfr.org/blog/russian-disinformation-popularizes-sputnik-v-vaccine-africa>

[27] <https://sputnikvaccine.com/newsroom/pressreleases/sputnik-v-team-statement-on-fake-news-in-uk-media/>

[28] <https://www.eeas.europa.eu/eeas/building-immunity-disinformation_en>

[29] <https://www.ncbi.nlm.nih.gov/pmc/articles/PMC8115834/>

[30] <https://www.comminit.com/polio/content/countering-online-vaccine-misinformation-eueea>

of Sputnik V while criticizing Western vaccines.[31] Ukraine, where Russian outlets spread contradictory and confusing vaccine narratives. Baltic states, where Russian media falsely claimed NATO troops were spreading COVID-19. The report on "Countering online vaccine misinformation in the EU/EEA" by the European Centre for Disease Prevention and Control (ECDC) (2021) highlights the findings of social media analysis and stakeholder consultations across six case study countries, focusing on vaccine misinformation. The proportion of misinformation in these countries ranged from 3 percent (Spain) to 12 percent (Netherlands and Romania), with COVID-19 being the most frequently discussed disease, accounting for 68 percent of all posts. Although misinformation rates were similar across diseases, the high volume of COVID-19-related content resulted in a greater overall spread of false information.

In France, Germany, and the Netherlands, a few key sources repeatedly disseminated misinformation, some reaching wide audiences through platforms like YouTube and Twitter.[32]

The campaigns used various tactics such as coordinated networks of fake social media accounts and websites, state-backed media outlets like Sputnik, impersonation of legitimate news sources, exploitation of real events and concerns to spread misleading information.[33]

EU institutions and member states worked to counter these narratives through fact-checking, media literacy efforts, and coordinated strategic communications.[34] However, the disinformation campaigns remained a significant challenge throughout the pandemic period.

Fabricated Stories About Ukrainian Refugees Committing Crimes in Poland and in other EU Countries (2022–2023)

False stories about Ukrainian refugees committing crimes have been circulating widely across multiple EU countries since the start of the Russian

---

[31] <https://misinforeview.hks.harvard.edu/article/clarity-for-friends-confusion-for-foes-russian-vaccine-propaganda-in-ukraine-and-serbia/>
[32] <https://www.ecdc.europa.eu/sites/default/files/documents/Countering-online-vaccine-misinformation-in-the-EU-EEA.pdf>
[33] <https://www.nato.int/cps/en/natohq/177273.htm>
[34] <https://www.pubaffairsbruxelles.eu/eu-institution-news/vaccine-disinformation-platforms-monitoring-program-extended-until-the-end-of-the-year/>

invasion in 2022. These campaigns aim to stir up anti-refugee sentiment and undermine support for Ukraine.[35]

Typical false claims include allegations of Ukrainians engaging in violent crimes, theft, and ungrateful or entitled behavior. These stories often portray refugees as dangerous or a burden on host countries.[36]

Many of these fabricated stories spread rapidly on social media platforms like Facebook, Twitter, TikTok and Telegram. They often use out-of-context images or videos to make false claims seem more credible. While these false narratives have appeared across the EU, they have been particularly prevalent in countries hosting large numbers of Ukrainian refugees, such as Poland, Germany, and the Czech Republic.[37]

EU officials and researchers have identified Russian state actors and pro-Kremlin networks as key sources of these disinformation campaigns, viewing them as part of a broader strategy to undermine European unity and support for Ukraine.[38]

Fact-checkers and government agencies in various EU countries have been actively working to identify and debunk these false stories.[39] However, the volume and speed of disinformation often outpace these efforts. While the majority of these stories are fabricated, they can have real consequences by influencing public opinion and potentially affecting policy decisions regarding refugee support.

The Doppelgänger Campaign

The "Doppelgänger" campaign, a sophisticated Russian disinformation operation established in 2022, has been targeting multiple European countries

---

[35] <https://www.rferl.org/a/ukraine-eu-refugees-fatigue/32730367.html>

[36] <https://www.isdglobal.org/digital_dispatches/a-false-picture-for-many-audiences-how-russian-language-pro-kremlin-telegram-channels-spread-propaganda-and-disinformation-about-refugees-from-ukraine/>

[37] <https://www.isdglobal.org/digital_dispatches/a-false-picture-for-many-audiences-how-russian-language-pro-kremlin-telegram-channels-spread-propaganda-and-disinformation-about-refugees-from-ukraine/>

[38] <https://uacrisis.org/en/ukrayinski-bizhentsi-v-yes-u-fokusi-rosijskoyi-propagandy>

[39] <https://edmo.eu/publications/disinformers-use-similar-arguments-and-techniques-to-steer-hate-against-migrants-from-ukraine-or-the-global-south-2/>

and the United States by creating fake versions of legitimate news outlets to spread pro-Russian narratives.

First identified in 2022, the Doppelgänger campaign is believed to be operated by Russian entities, including the Russian Social Design Agency and Structura National Technologies.[40] The primary objectives of the campaign was to a) Undermine Western support for Ukraine, b) Promote pro-Russian narratives about the war, c) Sow discord among European countries and d) Exploit political and social vulnerabilities in target nations

The Doppelgänger campaign employed several sophisticated tactics:

1. Website Cloning: The operation creates near-perfect replicas of legitimate news websites, including major outlets like Der Spiegel, The Guardian, Le Monde, and The Washington Post.[41]
2. Domain Spoofing: Operators purchase domain names similar to authentic media outlets, often using alternative top-level domains (e.g., .ltd, .online).[42]

Using websites like "nato[.]ws" and "RRN[.]media," the campaign falsely claims NATO is doubling its military budget and exaggerates the impact of Ukrainian refugees in Europe. Bots and social media platforms like Facebook and Twitter amplify the content, and paid posts increase visibility. This cross-platform operation blurs the lines between fact and fiction, targeting public opinion with manipulated media.[43]

3. Content Manipulation: The fake sites publish a mix of genuine news and fabricated stories aligned with pro-Russian narratives[44]—The Doppelgänger disinformation campaign leveraged generative AI and

---

[40] <https://en.wikipedia.org/wiki/Doppelganger_(disinformation_campaign)>
[41] <https://www.doppel.com/blog/russian-disinformation-campaign-doppelganger-is-why-doppel-exists>
[42] <https://www.cybercom.mil/Media/News/Article/3895345/russian-disinformation-campaign-doppelgnger-unmasked-a-web-of-deception/>
[43] <https://www.cybercom.mil/Media/News/Article/3895345/russian-disinformation-campaign-doppelgnger-unmasked-a-web-of-deception/>
[44] <https://www.doppel.com/blog/russian-disinformation-campaign-doppelganger-is-why-doppel-exists>

fake domains mimicking legitimate media outlets like *Bild* and *The Guardian* to spread pro-Russian narratives.[45]

4. Social Media Amplification: The campaign uses networks of inauthentic social media accounts to spread links to the fake articles[46]—"The Counter Disinformation Network (CDN)," a collective of 130 professional fact-checkers led by non-profit Alliance4Europe, found over 1,300 pro-Russian posts from June 4 to 28 resembling the Russian Doppelgänger campaign, a targeted disinformation operation that "relies on imitating legitimate media entities" to spread "narratives beneficial to Russia."[47]

5. Targeted Advertising: Doppelgänger has purchased ads on platforms like Facebook to increase the reach of its content.[48]

6. Multi-language Approach: Content was produced in multiple European languages, including English, German, French, Italian, and Polish.[49]

Disinformation on Energy and Climate

Efforts to transition from fossil fuels face increasing disinformation campaigns. These campaigns aim to deny climate change and derail investments in renewable energy. Similarly, the fossil fuel industry has been accused of spreading misinformation to slow the adoption of renewables, often using social media to create echo chambers and exploit divisions.

The main discourses and mechanisms identified are as follows:

Denial of climate change or minimization of human impact: This mechanism involves denying climate change's existence or downplaying

---

[45] <https://www.cybercom.mil/Media/News/Article/3895345/russian-disinformation-campaign-doppelgnger-unmasked-a-web-of-deception/>

[46] <https://www.euronews.com/next/2024/09/05/threat-is-ongoing-as-russian-doppel-ganger-operation-continues-on-x-and-meta-despite-eu-pro>

[47] <https://www.euronews.com/next/2024/09/05/threat-is-ongoing-as-russian-doppel-ganger-operation-continues-on-x-and-meta-despite-eu-pro>

[48] <https://euvsdisinfo.eu/doppelganger-strikes-back-unveiling-fimi-activities-targe-ting-european-parliament-elections/>

[49] <https://www.ecpmf.eu/actions-must-be-taken-to-address-mass-pro-russian-spoofing-of-legitimate-media-outlets/>

humanity's role in causing it. False claims typically assert that GHG emissions have minimal impact on global warming or that temperature rises are solely due to natural variations. Such misinformation spreads across social media platforms through manipulated graphics and misleading scientific quotations. These narratives extend beyond fringe groups to influential voices who seek to maintain the status quo and block climate-related political action.

Media accused of climate alarmism: This discourse claims that mainstream media and environmental organizations exaggerate climate change effects to create unnecessary public panic. Viral articles falsely accuse major European media outlets of overstating Arctic ice melt rates. Despite substantial evidence from reliable institutions like NASA and the European Space Agency (ESA), these claims attempt to undermine legitimate scientific reports. By questioning media reliability, this misinformation aims to reduce public willingness to address climate issues.

Discrediting activists: This type of disinformation portrays climate activists as hypocrites by focusing on their personal behaviors, particularly their travel habits. Online messages often target prominent climate advocates, criticizing their carbon footprints while ignoring the systemic changes they promote. This approach diverts attention from broader policy reforms and undermines public trust in the climate movement. By highlighting perceived personal inconsistencies, it aims to weaken support for broader environmental and social justice goals.

Link to wider conspiracy theories: Climate misinformation often connects to broader conspiracy theories. These narratives portray global climate initiatives as schemes by elites to impose authoritarian control. Such conspiracy theories, which frame climate action as part of a larger plot by global elites, spread through sensationalist social networks.

NATO's 2024 report found Kremlin-backed actors actively pushing climate change denial and opposing renewable energy investments, aligning with Russia's interest in keeping Europe dependent on gas. Efforts to promote renewables and electric heat pumps reduce this dependence and limit

Putin's influence.[50] The report highlights foreign disinformation campaigns targeting clean energy transitions, misdirecting focus toward gas reliance. Renewables like wind and solar are key to reducing foreign energy vulnerability, as reliance on gas has significantly increased energy costs since the Russian invasion of Ukraine.[51]

This is especially so in the case of wind turbines, a key symbol of renewable energy, have become the target of disinformation narratives, particularly in northern Europe. False claims, such as wind farms releasing harmful chemicals, have circulated widely. These baseless stories often exaggerate environmental impacts and are amplified through social media to undermine support for green energy. Recent examples include fabricated stories about wind turbines being responsible for reindeer deformities in Norway and false links between wind farms and PFAS (Perfluoroalkyl and Polyfluoroalkyl Substances) water contamination in Denmark.

These narratives reflect broader disinformation efforts to stall the transition to sustainable energy sources.[52]

IV. Social Media Manipulation

Social media manipulation has become a critical tool for spreading disinformation, amplifying narratives, and influencing public opinion. This section explores several instances of social media manipulation, focusing on tactics such as fake accounts, bot networks, and coordinated inauthentic behavior.

One example is how foreign-linked accounts amplified anti-EU sentiment during the 2016 Brexit referendum. These efforts included the widespread use of social media bots and coordinated campaigns to sway public perception. Another example was the use of coordinated inauthentic behavior targeting the 2019 EU Parliamentary Elections, involving fake accounts and misleading content to influence voters in multiple EU countries. These examples reveal how social media manipulation can amplify disinformation, destabilize

---

[50] <https://news.sky.com/story/campaigns-of-misinformation-around-heat-pumps-says-energy-minister-amid-record-number-of-installations-13052428>

[51] <https://eciu.net/media/press-releases/2024/nato-kremlin-backed-actors-disinformation-seeks-to-derail-green-investment-comment>

[52] <https://edmo.eu/publications/wind-turbines-and-poisoned-animals-a-new-denials-popular-disinformation-narrative-against-renewable-energy/>

societies, and influence political outcomes. Understanding these tactics is crucial for developing effective strategies to counter disinformation and protect democratic processes.

Manipulation During the Brexit Referendum

Evidence suggests that social media accounts played a role in amplifying anti-EU messaging during the 2016 Brexit referendum campaign. A study by 89up, a digital agency, found that Kremlin-backed outlets RT and Sputnik had more reach on Twitter for anti-EU content than either the official Vote Leave or Leave.EU campaigns. The report estimated that Russian media interference in the referendum was worth up to £4 million in terms of social media reach.[53]

Twitter later revealed that Russian-linked accounts sent over 10 million tweets in an effort to spread disinformation and discord, including a significant push on the day of the Brexit vote. On June 23, 2016, Russia mobilized an army of trolls that at one point included 3,800 accounts, which tweeted out 1,102 posts with the hashtag #ReasonsToLeaveEU.[54] Overall, Russian-linked accounts tweeted the phrase "Brexit" more than 4,400 times during their period of activity, although mostly after the referendum had taken place.[55,56]

However, the exact impact of these efforts remains difficult to quantify. The UK Parliament's Intelligence and Security Committee report, released in 2020, criticized the government for failing to properly investigate potential Russian interference in the referendum. The report stated that no serious effort was made to examine Russian attempts to influence the vote, describing this lack of investigation as "astonishing."[57]

Hate Narratives

Disinformation campaigns targeting EU politicians and institutions have become a growing concern during the European Elections held in June 2024. These campaigns often focus on smear campaigns against European

---

53 <https://89up.org/russia-report>
54 <https://www.telegraph.co.uk/technology/2018/10/17/russian-iranian-twitter-trolls-sent-10-million-tweets-fake-news/>
55 <https://www.bbc.com/news/technology-45894486>
56 <https://www.telegraph.co.uk/technology/2018/10/17/russian-iranian-twitter-trolls-sent-10-million-tweets-fake-news/>
57 <https://foreignpolicy.com/2020/07/21/britain-report-russian-interference-brexit/>

Politicians while promoting anti-Ukrainian narratives, anti-EU themes, and pro-Russian content.

Further, Global Disinformation Index (GDI) has observed a rise in online gendered abuse and disinformation targeting female EU leaders and candidates, including President Ursula von der Leyen. These narratives often include elements of misogyny, racism, homophobia, and foreign interference, with von der Leyen facing particularly high levels of attacks. The disinformation portrays women as incompetent and manipulative, frequently using misogynistic language. Female politicians of color face additional racist abuse. Russian-backed media has also contributed to these narratives, particularly in the context of the Ukraine war. These attacks undermine women's participation in public life.[58,59]

Disinformation about the 2025 Wildfires in Los Angeles California

The wildfires spreading around Los Angeles California in January 2025, rank among the most devastating in the city's history, leaving countless residents grappling with the loss of their homes and communities. Amid this tragedy, social media platforms have become fertile ground for misleading and conspiratorial content. Many accounts, often promoting climate skepticism or alternative, unsubstantiated explanations for the fires, have gained traction.

Platforms such as Instagram and Telegram have been hotspots for these narratives, with accounts like "Climat Realist," "JakeGTV," and "MTRXISREAL Uncensored" playing a central role in circulating speculative or fabricated content. These posts often feature narratives rooted in climate denial or conspiracies. For instance, a blurry video shared by JakeGTV on Instagram purportedly shows individuals dressed in black, with the accompanying narration suggesting they are either "vandals or federal agents" deliberately setting the fires. The video, however, lacks verifiable information about its origin, location, or the individuals depicted, and no credible evidence supports these claims. To further engage viewers, the post included a poll asking them to speculate on whether the individuals were vandals or agents of the federal government, fueling distrust and unfounded narratives.

---

[58] <https://ec.europa.eu/newsroom/edmo/newsletter-archives/53846>
[59] <https://www.disinformationindex.org/blog/2024-06-10-gendered-disinformation-in-the-european-parliamentary-elections/>

A recurring claim in several posts focuses on observations from wildfire-affected areas, noting that while many homes have been destroyed, nearby trees would have remained unburned. Although this pattern is frequently cited in wildfire conspiracy theories, it is rarely accompanied by scientific context or analysis. Some users have linked this phenomenon to other fires, such as those in Hawaii, alleging, without evidence, that these events are part of a deliberate government scheme. For example, JakeGTV shared an image comparing the Los Angeles wildfires to photos from Maui and Paradise, highlighting intact trees amidst widespread destruction to suggest the existence of "abnormal fires" as part of a coordinated plan.

Others have drawn connections to recent events, such as the cancelation of 72,000 homeowners' fire insurance policies in California earlier in 2024. Posts argue that this was not a coincidence, implying that the cancelations were part of a larger orchestrated effort to leave residents unprotected before the fires occurred. While the insurance cancelations are a documented fact, these posts use them to bolster conspiracy theories suggesting a premeditated agenda behind the wildfires.

In contrast, some posts focus less on elaborate conspiracies and more on denying the link between wildfires and climate change. For instance, a post on the Telegram channel "Climat Realist" outright dismissed the idea that climate change played any role in the fires, denying both current impacts and future risks associated with global warming.

Cross Platforms Manipulation

One of the key methods of disinformation through platforms' manipulation involves the use of multiple platforms, such as Telegram, VKontakte, and X (Twitter), to spread anti-EU and anti-Ukraine narratives. Russian-linked campaigns (but not only) have been particularly active, employing coordinated strategies to disseminate propaganda across social media channels and maintain reach even after bans from major platforms. This tactic highlights how different platforms are leveraged to sustain influence and continue spreading manipulated content.

Another example is the disinformation targeting the European Green Deal, where campaigns are using Facebook, Twitter, and other platforms to spread misleading claims about climate policies. These campaigns often link climate action to unrelated controversial topics, such as COVID-19, and exaggerate the negative impact of the Green Deal to undermine public support.

Additionally, false narratives regarding European energy independence and the transition to renewable energy are amplified across platforms to create confusion and resist change.

Cross-platform coordination enables disinformation campaigns to spread seamlessly across different channels, increasing their impact and making them more challenging to counter. Addressing these campaigns requires coordinated responses, transparency from social media platforms, and effective fact-checking to limit the spread of manipulated content.

Disinformation about the USAID Shutdown

Under the pretext of reducing the U.S. budget deficit, the Trump administration, in coordination with the unofficial Department of Government Expenditure (DOGE) led by Elon Musk, has taken steps to dismantle USAID early 2025—placing almost all employees on leave and freezing foreign aid programs. This move has thrown billions of dollars in foreign aid into uncertainty, effectively stalling vital humanitarian initiatives.

The shutdown of USAID was justified through a campaign of misleading claims, exaggerations, and outright fabrications. The administration portrayed the agency as inefficient, wasteful, and ideologically biased, using selective data points to construct a narrative of dysfunction. Some of the most widely circulated claims included

- $2 million for Moroccan pottery classes—In reality, this funding supported Moroccan artisans in preserving cultural heritage and expanding economic opportunities.
- $40 billion for electric vehicle (EV) ports with only eight completed— The actual allocation was $7.5 billion, with over 200 chargers already operational and thousands more in development.
- $1.5 million to advance DEI (Diversity, Equity, and Inclusion) initiatives in Serbian workplaces—A mischaracterization of broader economic and governance initiatives aimed at promoting workplace equity.
- $50 million taxpayer dollars that went out the door to fund condoms in Gaza— There are no records or evidence of such spending. The administration referenced a $102 million USAID grant to the International Medical Corps (IMC), a non-governmental organization (NGO) providing medical aid in Gaza. IMC clarified that since the

October 7 Hamas attack, it had received over $68 million from USAID but stated that no U.S. funding was used for condoms or family planning.
· $70,000 for a DEI musical in Ireland, $47,000 for a trans opera in Colombia, and $32,000 for a trans comic book in Peru—These figures were cherry-picked to sensationalize minor expenditures within much larger cultural diplomacy programs.

These distortions were not mere mistakes, but deliberate strategies designed to evoke outrage and justify a sweeping policy shift. This tactic aligns with a broader trend in contemporary American politics: using disinformation to undermine institutions and consolidate power.

The disinformation campaign extended beyond domestic sources. A Russian disinformation operation circulated a fabricated report falsely attributed to the U.S. broadcaster E!. This report claimed that celebrities such as the movie actors Ben Stiller and Angelina Jolie had received USAID funds for visits to Ukraine to bolster President Zelensky's public image. E! denied producing the report, and verifiable records confirm that neither of them received any government funding. Both financed their trips independently.

What is particularly significant is not merely that Russian media propagated this false narrative, but that Elon Musk and Donald Trump Jr. actively amplified it, sharing and endorsing the disinformation as factual. While fact-checkers suggested they "fell for it," a more precise interpretation is that they were indifferent to its accuracy. This was not a simple misstep but rather a strategic effort to undermine a U.S. federal department with disinformation to gain public support for their political agenda. The administration's alignment with Russian disinformation actors makes the United States more vulnerable to disinformation threats when the presidential apparatus itself amplifies false stories. This alliance will likely continue whenever the Trump administration seeks to undermine opposition or institutional obstacles to consolidating power.

V. Coordinated Inauthentic Behavior and Interference in Elections

Coordinated Inauthentic Behavior (CIB) is a manipulative tactic that uses fake, authentic, and duplicated social media accounts to spread disinformation and manipulate public opinion. CIB campaigns can be harmful and threaten democracy and freedom of expression.[60] In effect, it operates as an adversarial network (AN) across multiple social media platforms.[61] CIB operations can include psychological harassment, exploiting technology features and logic, attacking opponents and unaware targets, and spreading disinformation.[62]

In February 2019, just before Moldova's parliamentary elections, Facebook removed more than 100 accounts and pages for engaging in CIB in Moldova.[63]

In the lead-up to the May 2019 European Parliament elections, Facebook detected and removed several networks of accounts, pages, and groups engaged in coordinated inauthentic behavior targeting EU member states. Facebook set up an operations center in Dublin to coordinate rapid response to election-related issues across the EU.[64] The company reported taking down over 500 million fake accounts in the first three months of 2019 alone, many before they became active.

This coordinated inauthentic behavior and the interference in elections are not always detected and can become a threat in the future for very targeted and invisible interference. It is also worth noting that most of the detected foreign interference is indirectly linked to Russia, and there is a long history of disinformation dating back from the Cold War and the former Soviet Union. However, foreign interference from other governments and malicious groups does exist, notably from China, but also Israel, North Korea, Iran or from Islamic groups (former Daesh). It is expected that foreign interference in the future will not only be the fact of governments but of groups or opinion leaders (e.g., far-right movements) supported by several platforms.

Finally, there have been a lot of disinformation campaigns targeting international organizations to decrease their credibility such as the: Disinformation campaigns targeting the credibility of the Organisation for the Prohibition

---

60 <https://www.vicarius.io/vsociety/posts/understanding-coordinated-inauthentic-be-havior-cib-what-it-is-and-how-it-impacts-the-general-public>
61 <https://www.ncbi.nlm.nih.gov/pmc/articles/PMC10060790/>
62 <https://www.ncbi.nlm.nih.gov/pmc/articles/PMC10060790/>
63 <https://freedomhouse.org/article/together-we-are-stronger-social-media-companies-civil-society-and-fight-against>
64 <https://www.politico.eu/article/facebook-european-election-war-room-dublin-polit-ical-advertising-misinformation-mark-zuckerberg/>

of Chemical Weapons (OPCW). These campaigns, aimed to undermine the OPCW's findings and sow doubt about chemical weapons attacks, particularly in Syria. These campaigns are also likely to have a geopolitical impact when they are widespread, as they can influence the participants of a trial, or of an investigation.

VI. Remaining Challenges

The current situation faces persistent challenges in combating disinformation and misinformation, primarily due to the complexity of processes, threats, mechanisms, and divided responsibilities which cannot be clearly identified. (Gaborit 2024).[65]

Current debates about social media regulation overlook the role of malicious actors in spreading disinformation. Identifying these actors and tracing the origins of such threats is exceptionally challenging. Studies by NATO Stratcom and the EU Disinfo Lab demonstrate that "inauthentic coordinated behavior" is being used to spread fake news (Romero-Vicente et al. 2024). This complicates the identification of threats and makes it difficult to hold perpetrators accountable for their actions (Sessa et al. 2024). Without proper accountability mechanisms, a culture of impunity may emerge.

Disinformation tactics have proven to be highly adaptable. For example, on platforms like TikTok, the strategy involves attracting a young and liberal audience with engaging content before gradually introducing propaganda narratives once a sense of familiarity has been established and users are hooked. Fake news resonates with specific demographic groups, posing an increasing threat, particularly on platforms with weak protections, such as TikTok and X (formerly Twitter).

While pro-Kremlin media and channels have amplified disinformation about the war in Ukraine, this has not prevented the spread of false information on other topics, such as climate change, where the actors behind the narratives are less easily identifiable. Further research is needed in this area.

---

[65] You can also find these discussion points in *in my recent article* A sociological Approach to Disinformation and AI: concerns, responses and challenges *in Journal of Political Science and International Relations*, 2024, Vol. 7, n°4, 75–88 <https://doi.org/10.11648/j.jpsir.20240704.1>

The lines between "debunking" disinformation and its detection by security agencies are also blurred. Large-scale manipulative disinformation campaigns are still primarily uncovered by cybersecurity agencies. For instance, the French agency **Viginum** was established in 2021 to detect digital interference by foreign entities seeking to influence public opinion.

In February 2024, the agency informed the media that it had identified over 193 websites spreading disinformation via social media and messaging apps. According to Viginum, the disinformation campaign amplified conspiracy theories and fueled tensions. Even for security agencies, pinpointing the exact origin of such campaigns is not always straightforward, particularly when misinformation or disinformation must be traced back to foreign governments or individual actors acting in their country's interest.

Furthermore, hybrid warfare, in which disinformation is just one tool, combines cyberattacks and large-scale disinformation, posing serious risks of malicious influence not only on the media, governments, and public infrastructure but also on civil society and academic institutions.

Social media platforms are moving away from traditional fact-checking, opting instead for community-based systems like Community Notes that rely on user input. This shift raises concerns about fighting misinformation effectively at scale. Independent fact-checking has grown substantially—doubling to nearly 400 teams of journalists and researchers across 105 countries in six years—but this growth has slowed, and measuring its impact remains challenging. Fact-checkers struggle with limited resources, reach, and their reactive approach, as they often address false information only after it has spread.

Major platforms and European media outlets have created their own fact-checking teams, but these can't keep pace with the rapid spread of false information and coordinated disinformation campaigns. While fact-checkers play a vital role in promoting critical thinking and teaching people to verify information, they often can't prevent false narratives from going viral, especially against organized campaigns. The "Matryoshka" campaign of January 2024 exemplifies this challenge—it specifically targeted fact-checkers to mock their work and erode public trust.

The fight against disinformation becomes even more challenging as platforms reduce content moderation and institutions scale back fact-checking efforts. Despite their importance, fact-checkers alone cannot keep up with the scale and complexity of disinformation, underscoring the need for stronger **preventive measures** to protect public discourse.

Research indicates that *"hostile actors continue to develop innovative strategies to bypass disinformation-blocking mechanisms"* (Bergmanis-Koräts et al., 2024: 9). Independent journalism, with its ability to cross-check information, has never been more vital than in recent years. However, technological shifts threaten their economic models (Lloyd et al., 2015; Kunelius et al., 2027; Newman et al., 2024), making it increasingly difficult for traditional media to secure sustainable funding mechanisms.

Interestingly, the platform Google has created a "prebunking" initiative aiming at empowering individuals to spot, prevent and detect disinformation online.[66] The platform helps to detect 11 manipulation techniques: Emotional language, false dichotomy, cherry-picking, fake experts, red herring, scapegoating, ad hominem, polarization, impersonation, slippery slope, decontextualization. The 11 tactics are detailed in the box below.

1. **Emotional Language**: Using emotions and words charged with emotions to provoke strong feelings such as fear, sadness, empathy and anger to sway opinions.
2. **False Dichotomy**: Presenting only two options, one clearly unfavorable to the author(s) and the one of the author(s)—to force a particular choice (e.g., *there are the ones who share our views, and the others ...)*
3. **Cherry-Picking**: Selecting data that supports a specific viewpoint while ignoring contradictory evidence. This strategy is commonly employed in climate change disinformation campaigns.
4. **Fake Experts**: Citing individuals who appear authoritative but lack relevant expertise to validate false claims.
5. **Red Herring**: Introducing irrelevant information to distract from the main issue.
6. **Scapegoating**: Unjustly blaming a person or group for problems to deflect responsibility.
7. **Ad Hominem**: Attacking an opponent's character instead of addressing their arguments.
8. **Polarization**: Dividing people into opposing groups to create conflict and hinder unity.

---

[66] <https://prebunking.withgoogle.com/>

9. **Impersonation**: Mimicking trusted sources to deceive and spread false information.
10. **Slippery Slope**: Arguing that a minor action will lead to significant and often negative consequences without evidence.
11. **Decontextualization**: Presenting information without its full context to mislead audiences. Disinformation campaigns targeting the USAID agency are presented in a decontextualized manner. For example, describing its work as "support for pottery classes in Morocco" fails to acknowledge its role in supporting job-creating economic activities. Similarly, mentioning millions spent for the "shipment of condoms" to other countries without providing the context of critical HIV/AIDS prevention makes it seem completely absurd and ridiculous.

Box 1: Tactics identified by the Google Pre Bunking initiative[67]

These tactics can indeed be used to identify disinformation, but not to prevent its large-scale spread, as the use of fake accounts and bots facilitates the dissemination of disinformation campaigns, making them difficult to block.

Another impact, which should not be overlooked, is the impact of targeted disinformation on election results in democratic countries (Wade 2018). For this, preventive, and not only reactive measures should be taken as responses.

By leveraging the collection and manipulation of user data to anticipate and influence voters' political opinions and election outcomes, user profiling and micro-targeting can pose a serious threat to democracy, public debate, and voter decision-making (Kertysova, 2018; Mont'Alverne et al., 2024).

This issue is particularly crucial, as 2024 was an election year for half of the world's population, yet concerns over electoral interference remain unresolved. Evidence is difficult to trace, and even with AI-driven tools, current detection mechanisms are still not fully capable of exposing all fake news.

Finally, while user profiling and political micro-targeting may seem like commercial strategies, these practices also pose issues regarding privacy and personal data protection. This takes us back to the underlying process of amassing and processing of vast amounts of personal data which is used in AI systems.[68] "Such data can be […] stripped of its original purpose(s) and

---

[67] <https://prebunking.withgoogle.com/>
[68] and which will also be used in the AI related debunking systems

may be used for objectives the individual is largely unaware of—in this case, profiling and targeting with political messages—in contravention of existing EU data protection principles" (Kertysova 2018).[69]

The increasing level of fake news is likely to have an impact by creating doubts, confusion and concerns among the readers, with the effect to dilute the information on true facts. There are difficulties and challenges in identifying up front what can be considered as trustworthy information and what is not. The results of our research show that while disinformation is spreading and is being used in increasingly complex and aggressive campaigns, the existing responses are only at an early stage. The current available responses are indeed "reactive" and rely on the platforms for moderations and on the users of social media and online platforms to become more "critical" toward what they read. In addition to this, the authors of disinformation being in foreign countries, the questions of "accountability" and "transparency" are limited, as it is complex for public authorities to locate the authors of disinformation campaigns and to hold them accountable. In this context the current regulations, although comprehensive, cannot be efficiently enforced.

Furthermore, the repeated dissemination of disinformation by the Trump administration, and particularly by Elon Musk, raises serious concerns about governance rooted in falsehoods. Disinformation surrounding the potential dismantling of USAID could serve as a real-world test case to assess the extent to which disinformation can be used as a tool to weaken or dismantle institutions.

If this strategy proves effective here, what will prevent the administration from employing similar tactics to bypass Congress, disregard judicial rulings, or undermine the legitimacy of elections?

By normalizing disinformation as a means of governance, this administration is laying the groundwork for an unprecedented concentration of power, leveraging communication strategies that border on propaganda. While such practices are not new in history, what makes this particularly alarming is its connection to foreign interference and its amplification through cutting-edge technological tools. This combination creates an especially dangerous dynamic—one that threatens not only American democracy and the rule of law but also the fragile balance of global stability.

---

[69] Systems also need large training datasets to improve in accuracy and performance

Conclusion

Through this chapter, we have gained an initial insight into the threat posed by disinformation, analyzing its impact on society and the limitations of the proposed responses. Our analysis highlights a general context of impunity for malicious actors, despite cases of identity theft that exceed legal boundaries. It confirms the growing use of fake news and disinformation as political weapons in democratic debates, emphasizing the need for deeper research and effective preventive measures.

New technologies create opportunities for online communities and collective thinking. However, they also contribute to the spread of disinformation, indirectly leading to a collapse of authority and an erosion of values. Traditional authority figures are being replaced by unlimited access to information and connectivity, including manipulated and falsified content. Civil society, the private sector, and leaders—including businesses, communities, and political figures—must address these challenges. According to security experts (Kueng, 2020), future leaders will face a new environment characterized by widespread uncertainty, the need for versatility, the ability to build new communities, and increased support for collective action.

The digital revolution is undoubtedly shaping a world of uncertainty, causing upheavals in economy, politics, and lifestyles. Young generations must adapt to these rapid changes by developing IT skills, media literacy, and an understanding of algorithmic processes. Leaders, freelancers, students, and employees must cultivate new skills such as e-governance, technological proficiency, communication abilities, and the capacity to manage an overwhelming flow of instant information. However, limitations are necessary in order to prevent technological advancements—such as the rise of algorithms or robots—from dictating our ways of life.

Disinformation is a shared concern for civil society. The digital revolution—accelerated by AI but rooted in earlier innovations—is a key force shaping the future. Our research findings indicate that responses to disinformation in European countries remain ineffective, despite the existence of a robust regulatory framework. As this chapter illustrates, we are witnessing a major shift in how we access information, while existing responses remain in their infancy.

Our analysis also reveals that European and other countries are increasingly aware of the impact of disinformation. However, it also highlights a gap between the ability to identify fake news and disinformation, and the responses implemented. Moreover, there is a limited understanding of the strategies, threats, and actors involved in its dissemination.

But imagine a world where there are no facts, no truth, no science ... A world of lies and manipulation.

References

AI4DEBUNK 2024, "Towards of Theory Framework," AI4debunk, Riga, March 13, 2024.

Antoniuk, D. "Russian 'influence-for-hire' firms spread propaganda in Latin America: US State Department", The Record by Recorded Future, November 8, 2023. <https://therecord.media/russia-influence-for-hire-firms-latin-america-propaganda-us-state-department>

Art, S. "Media literacy and critical thinking." *International Journal of Media and Information Literacy, 3*(2), 2018, 66–71.

Bauer M., Cahlíková J., Chytilová J., Roland G., Želinský T. "Shifting Punishment onto Minorities: Experimental Evidence of Scapegoating," *The Economic Journal*, Volume 133, Issue 652, May 2023, 1626–1640, <https://doi.org/10.1093/ej/uead005>

Bergmanis-Koräts G., Arhippainen M. et al. "Virtual Manipulation Brief. Hijacking Reality. The increased role of Generative AI in Russian Propaganda", NATO Stratcom. 2024 <https://stratcomcoe.org/publications/virtual-manipulation-brief-20241-hijacking-reality-the-increased-role-of-generative-ai-in-russian-propaganda/307>

Betz H. G., Oswald M. L. "Emotional Mobilization: The affective Underpinnings of Right-Wing Populist Party Support", *Palgrave Handbook of Populism*, 2021, pp. 115–143.

Bollmann H. S., & Gibeon G."The spread of hacked materials on Twitter: A threat to democracy? A case study of the 2017 Macron Leaks" (Doctoral dissertation, Hertie School), 2022.

Bontridder N. and Poullet Y. "The role of artificial intelligence in disinformation", *Data & Policy*, 3: 2021, e32. <https://doi.org/10.1017/dap.2021.20>

Butcher P., & Neidhardt A. H. "Fear and lying in the EU: Fighting disinformation on migration with alternative narratives", Foundation for European Progressive Studies, 2020.

Charillon F. *Guerres d'influence*, Odile Jacob, 2018.

Cull. N. J. "Public Diplomacy: lessons from the past", USC center of public diplomacy, 2009.

Darwin, Rusdin D., Mukminatien N., Suryati N., Laksmi E. D., & Marzuki. "Critical thinking in the AI era: an exploration of EFL students' perceptions, benefits, and limitations. Cogent Education," 11(1), 2290342. 2024. <https://doi.org/10.1080/2331186X.2023.2290342Page%205%20of%2018>

Dauksas V., Venclauskiené L., Urbanaviciuté K., Friedman O. "War on all fronts: How the Kremlin's Media Ecosystem broadcasts the war in Ukraine." NATO Stratcom <https://stratcomcoe.org/publications/war-on-all-fronts-how-the-kremlins-media-ecosystem-broadcasts-the-war-in-ukraine/301>

Deutsch M. "Trust and Suspicion," *Conflict Resolution Number* 2, (Vol. 8), 1958.

European Union Agency for Cybersecurity (i), Lella I., Ciobanu C., Tsekmezoglou E. "ENISA threat landscape", 2023: July 2022 to June 2023. Retrieved from: <https://data.europa.eu/doi/10.2824/782573>

European Union Agency for Cybersecurity (ii), Tsekmezoglou, E., Lella, I., Malatras, A. et al., *"ENISA threat landscape for DoS attack" – January 2022 to August 2023*, European Union Agency for Cybersecurity, 2023, retrieved from: <https://data.europa.eu/doi/10.2824/859909>

European Commission "A Multi-dimensional Approach to Disinformation: Report of the Independent High-Level Group on Fake News and Online Disinformation." Directorate-General for Communication Networks, Content and Technology, 2018a. Available at <https:// ec.europa.eu/digital-single-market/en/news/final-report-high-level-expert-group-fake-news-and-online-disinformation>

European Commission "Code of Practice on Disinformation" 2018b. Available at <https://ec.europa.eu/digital-single-market/en/news/code-practice-disinformation>.

European Commission "Synopsis Report of the Public Consultation on Fake News and Online Disinformation" 2018c. <https://ec.europa.eu/

digital-single-market/en/news/synopsis-report-public-consultation-fake-news-and-online-disinformation>

European Commission "Tackling Online Disinformation: A European Approach" (Communication) COM, 2018d, 236 final. <https://eur-lex.europa.eu/legal-content/EN/TXT/?uri=CELEX%3A52018DC0236>

European Commission "Assessment of the Code of Practice on Disinformation – Achievements and areas for further improvement. Commission Staff working document" (SWD(2020) 180 final).2020a

European Commission "European Democracy Action Plan" (Communication) COM(2020) 790 final 2020b. <https://eur-lex.europa.eu/legalcontent/EN/TXT/?uri=COM%3A2020%3A790%3AFIN&qid=1607079662423>

European Commission "Guidance on Strengthening the Code of Practice on Disinformation" (COM(2021) 262 final).2021a, <https://digital-strategy.ec.europa.eu/en/library/guidance-strengthening-code-practice-disinformation>

European Commission "Proposal for a REGULATION OF THE EUROPEAN PARLIAMENT AND OF THE COUNCIL LAYING DOWN HARMONISED RULES ON ARTIFICIAL INTELLIGENCE (ARTIFICIAL INTELLIGENCE ACT) AND AMENDING CERTAIN UNION LEGISLATIVE ACTS, COM/2021/206 final," 2021b.

European Commission "COMMISSION DELEGATED REGULATION (EU) .../... of 20.10.2023 supplementing Regulation (EU) 2022/2065 of the European Parliament and of the Council, by laying down rules on the performance of audits for very large online platforms and very large online search engines, C(2023) 6807 final," 20/10/2023.

Fine G. A. "Rumor, Trust and Civil Society: Collective Memory and Cultures of Judgment", *Diogenes* 2007, 54 (1):5–18. <https://doi.org/10.1177/0392192107073432>

Foreign Affairs Review "The meaning of sharp power : How authoritarian States project Influence," *Foreign Affairs Review*, November 16, 2017.

Gaborit P. "A sociological Approach to Disinformation and AI: concerns, responses and challenges", *Journal of Political Science and International Relations*, 2024, Vol. 7, n°4, 75–88, 2024, <https://doi.org/10.11648/j.jpsir.20240704.11>

Gaborit P. *Restaurer la confiance après un conflit civil*, L'Harmattan, 2009a.

Gaborit P. : "La confiance après un conflit ou la confiance désenchantée," in Bertho A., Gaumont-Prat H. et Serry H. *Colloque international La confiance et le conflit*, Université Paris Vincennes Saint Denis, 2009b.

Girard R. *The Scapegoat*, Johns Hopkins University Press, 1986.

Goodhart D. *The future to somewhere: The populist revolt and the future of politics*. London, Hurst and Company, 2017.

Goodhart D. "The revenges of the places that don't matter (and what to do about it)", *Cambridge Journal of Regions, Economy and Society*, II(I), 2017: 189–201.

Grabner-Kräuter S. "Empiral Research in Online Trust. A Review and Critical Assessment", *International Journal of Human-Computer Study*.2003 <https://doi.org/10.1016/S1071-5819(03)00043-0>

Grinberg N., Joseph K., Friedland L., Swire-Thompson B., & Lazer D. "Fake news on Twitter during the 2016 US presidential election", *Science*, 363(6425), 2019, 374–378.

Haiduchyk T., Shevtsov A., Bergmanis-Koräts G. "AI in Precision Persuasion : Unveiling Tactics and Risks on Social Media." *NATO Stratcom 2024* <https://stratcomcoe.org/publications/ai-in-precision-persuasion-unveiling-tactics-and-risks-on-social-media/309>

Hamm J. A., van der Werff L., Osuna A. I., Blomqvist K., Blount-Hill K. L., Gillespie N., Tomlinson E. C. "Capturing the conversation of trust research", *Journal of Trust Research*, 14(1), 1–7, 2024 <https://doi.org/10.10 80/21515581.2024.2331285>

Hardin R. *Trust and Trusworthiness*. New York, Russel Sage foundation editions, series on trust, volume 4, 2002.

Hardin R. *Distrust*, NYC, Russell Sage Foundation, 2004.

Hersh M. A. "Barriers to ethical behaviour and stability: Stereotyping and scapegoating as pretexts for avoiding responsibility," *Annual Reviews in Control*, Volume 37, Issue 2, 2013, 365–381, <https://doi.org/10.1016/j.arcontrol.2013.f09.013>

Kertysova K. "Artificial Intelligence and Disinformation How AI Changes the Way Disinformation is Produced, Disseminated, and Can Be Countered," *Security and Human Rights* 29, 2018, 55–81.

King K., Wang B. "Diffusion of real versus misinformation during a crisis event: A big data driven approach", *International Journal of Information Management*, 71, 2023 <https://doi.org/10.1016/j.ijinfomgt.2021.102390>

Kueng L. "Hearts and Minds: Harnessing Leadership, Culture, and Talent to Really Go Digital," *Oxford University, Reuters Institute*, 2020.

Kunelius R., Heikkilä H., Russell A. and Yagodin D. (eds.) (published with I. B. Tauris). "Journalism and the NSA Revelations: Privacy, Security and the Press," 2017.

Lloyd J. and Toogood L. (published with I. B. Tauris). "Journalism and PR: News Media and Public Relations in the Digital Age", *Oxford University and Reuters institute*, 2015.

Luhmann, N. *Trust and Power: Two Works by Niklas Luhmann*. Translation of German originals Vertrauen 1968 and Macht 1975, Chichester: John Wiley, 1979.

Małecka, A. "Non-State Actors in Nation-State Cyber Operations", *Rocznik Bezpieczeństwa Międzynarodowego*, 18(1), 45–64, 2024.

Mont'Alverne C., Badrinathan S., Ross Arguedas A., Toff B., Fletcher R., and Kleis Nielsen R. "The Trust Gap: How and Why News on Digital Platforms Is Viewed More Sceptically Versus News in General," *Reuters Institute*, 2022 <https://reutersinstitute.politics.ox.ac.uk/trust-gap-how-and-why-news-digital-platforms-viewed-more-sceptically-versus-news-general>

Moravcsik A. "Taking preferences seriously: A Liberal Theory of International politics," *International Organization*, vol. 4, n°51, fall 1997, p. 513–533.

Newman N. "Digital News Project: Journalism, Media and Technology: Trends and Prediction", Oxford University, Reuters Institute, 2024.

Persily N. and Tucker J. A. *Social Media and Democracy The State of the Field, Prospects for Reform*, Cambridge University Press, 2021.

Putnam R. *Making Democracy Work: Civic traditions in Modern Italy*, Princeton University Press, 1993.

Romero Vincente A et al. "Coordinated Inauthentic Behavior", EU Disinfo Lab 2024.

Scott K. D. "The causal relationship between Trust and the assessed value of management by objectives", *Journal of management*, vol. 6, 1980, pp. 157–175.

Scheirer W. *A Review of a History of Fake Things on the Internet*, Stanford University Press, 2023 <http://www.sup.org/books/title/?id=35460>

Seligman A. *The problem of Trust*, Princeton, Princeton University Press, 1997.

Sessa M. G. "EU Disinfolab 'Connecting the Disinformation Dots'", Friedrich Nauman Foundation, 2023.

Sessa M. G. "Miguel R. The Doppelganger Case: Assessment of Platform Regulation on the EU Disinformation Environment", *NATO Stratcom*, 2024.

Shahbazi M., Bunker D. "Social media Trust: Fighting misinformation in the time of crisis", *Information Journal of Information Management*, 77, 2024. <https://doi.org/10.1016/j.ijinfomgt.2024.102780>

Six F. E., Latusek D. "Distrust: A critical review exploring a universal distrust sequence," *Journal of Trust Research*, 13:1, 1–23, 2024 <https://doi.org/10.1080/21515581.2023.2184376>

Smith R. B., Perry M., & Smith N. N. : "'Fake News' in ASEAN: Legislative responses", *Journal of ASEAN Studies*, 9(2), 2021. 117–137. <https://doi.org/10.21512/jas.v9i2.7506>

Smuha, Nathalie A. "Beyond the individual: governing AI's societal harm", *Internet Policy Review* 10.3 2021. <https://doi.org/10.14763/2021.3.1574>

Stahl B. C. "On the Difference or Equality of Information, Misinformation, and Disinformation: A Critical Research Perspective", *The International Journal of an Emerging Transdiscipline* • Volume 9, 2006, 083–096 <https://doi.org/10.28945/473>

Sztrompa P. *Trust a sociological theory*, New York, Cambridge University Press, 1999.

Tilly C. *Trust and Rule*, Cambridge University Press, 2005.

Walker C., Kalathil. S., Ludwig J. "The cutting edge of sharp Power," Journal of Democracy, 2020, 31(1) pp. 124–137.

Wade M. "Psychographics: The Behavioural Analysis That Helped Cambridge Analytica Know Voters' Minds", *The Conversation*, March 21, 2018, <https://theconversation.com/psychographics-the-behavioural-analysis-that-helped-cambridge-analytica-know-voters-minds-93675>

Whyte C. "Deepfake news: AI-enabled disinformation as a multi-level public policy challenge," *Journal of Cyber Policy*, 5:2, 2020; 199–217, <https://doi.org/10.1080/23738871.2020.1797135>

Witzel L. "5 Things You Must Know Now About the Coming EU AI Regulation," <https://medium.com/@loriaustex/5-things-you-must-know-now-about-the-coming-eu-ai-regulation-d2f8b4b2a4a9> 2021 pp. 128–146.

CHAPTER 4

Climate Action and SDGs Jeopardized in a Fragmented Environment

Imagine a world ravaged by storms, floods, and famines, where humanity has destroyed its greatest ally—the Earth.

Climate action is often encapsulated in the phrase "think global, act local." The urgency of the climate crisis demands a more nuanced approach: adaptive local strategies tailored to specific geographies, societies, and cultures, paired with bold global action to counter the widespread impacts of climate change. While some solutions exist—such as initiatives to protect critical global ecosystems like forests and oceans—our efforts remain insufficient to achieve true resilience to climate change.

Former U.S. President Barack Obama emphasized the need to not just "talk the talk but walk the walk" during his address to the 2016 North American Leaders Summit in Ottawa, Canada. This adage resonates strongly in the context of climate action. Transformative change requires moving beyond rhetoric to implement tangible, effective measures, despite the high costs and complex challenges involved.

A global framework for climate action was established in 2015 through the United Nations SDGs, which provide a comprehensive roadmap for reducing poverty, addressing hunger, improving access to education, and tackling climate change. These 17 interconnected goals represent humanity's best path to achieving a more equitable and sustainable world, essential for our survival. However, global cooperation on climate action has faced significant setbacks. The United States' withdrawal from the Paris Agreement, formalized in January 2025 by a presidential decree, marked a particularly troubling moment. This decision undermined international efforts to combat climate change, weakened global solidarity, and sent conflicting messages about the urgency of the crisis.

The research community unanimously agrees on the need for urgent measures to strengthen the resilience of societies, ecosystems, and territories against the impacts of climate change. Dialogue is a crucial tool in this effort, broadening the decision-making process and fostering interaction between policymakers and civil society. By opening minds, generating new ideas, and enabling the exchange of knowledge and resources, dialogue acts as a catalyst for action. However, dialogue alone is insufficient. Structural barriers such as conflicting interests, lack of funding, inconsistent policies, limited accountability, and political inaction continue to impede the full implementation of climate solutions.

To address these challenges, we must pursue bolder global decisions, design more coherent and actionable policies, and genuinely integrate climate impacts into governance frameworks. Building trust-based community networks is essential for ensuring the successful implementation of these strategies. Ultimately, achieving transformative climate action requires not only talking about solutions but also walking the walk through meaningful action. The future of our planet depends on the urgency, unity, and strength of our collective efforts. In this context, restoring trust is essential.

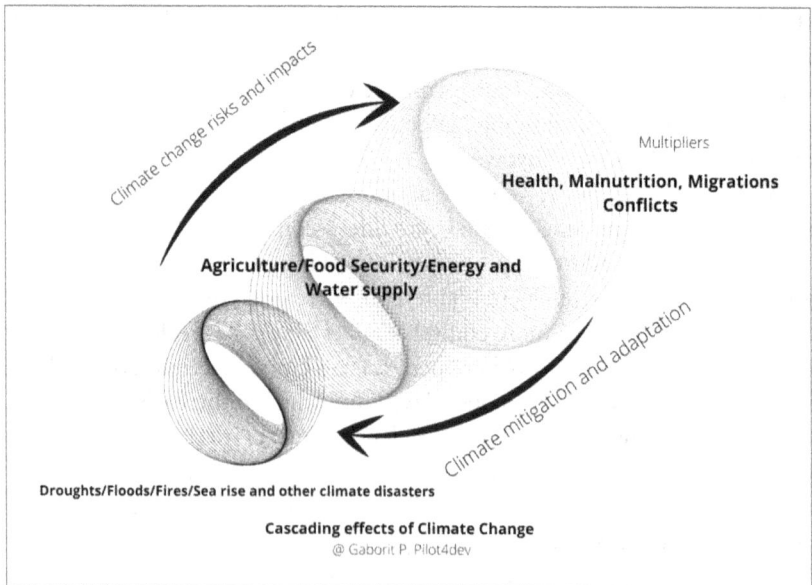

Figure 4. The impacts of climate change

I. The Sustainable Development Goals

The SDGs are a set of 17 global goals established by the United Nations in 2015 as part of the 2030 Agenda for Sustainable Development. These goals aim to address a wide range of global challenges, including poverty, inequality, environmental degradation, peace, and justice, while ensuring that no one is left behind.

The 17 SDGs:

1. **No Poverty**: End poverty in all its forms everywhere.
2. **Zero Hunger**: End hunger, achieve food security, improve nutrition, and promote sustainable agriculture.
3. **Good Health and Well-being**: Ensure healthy lives and promote well-being for all at all ages.
4. **Quality Education**: Ensure inclusive and equitable quality education and promote lifelong learning opportunities for all.
5. **Gender Equality**: Achieve gender equality and empower all women and girls.
6. **Clean Water and Sanitation**: Ensure availability and sustainable management of water and sanitation for all.
7. **Affordable and Clean Energy**: Ensure access to affordable, reliable, sustainable, and modern energy for all.
8. **Decent Work and Economic Growth**: Promote sustained, inclusive, and sustainable economic growth, full and productive employment, and decent work for all.
9. **Industry, Innovation, and Infrastructure**: Build resilient infrastructure, promote inclusive and sustainable industrialization, and foster innovation.
10. **Reduced Inequality**: Reduce inequality within and among countries.
11. **Sustainable Cities and Communities**: Make cities and human settlements inclusive, safe, resilient, and sustainable.
12. **Responsible Consumption and Production**: Ensure sustainable consumption and production patterns.
13. **Climate Action**: Take urgent action to combat climate change and its impacts.
14. **Life Below Water**: Conserve and sustainably use the oceans, seas, and marine resources for sustainable development.
15. **Life on Land**: Protect, restore, and promote sustainable use of terrestrial ecosystems, manage forests sustainably, combat desertification, and halt biodiversity loss.

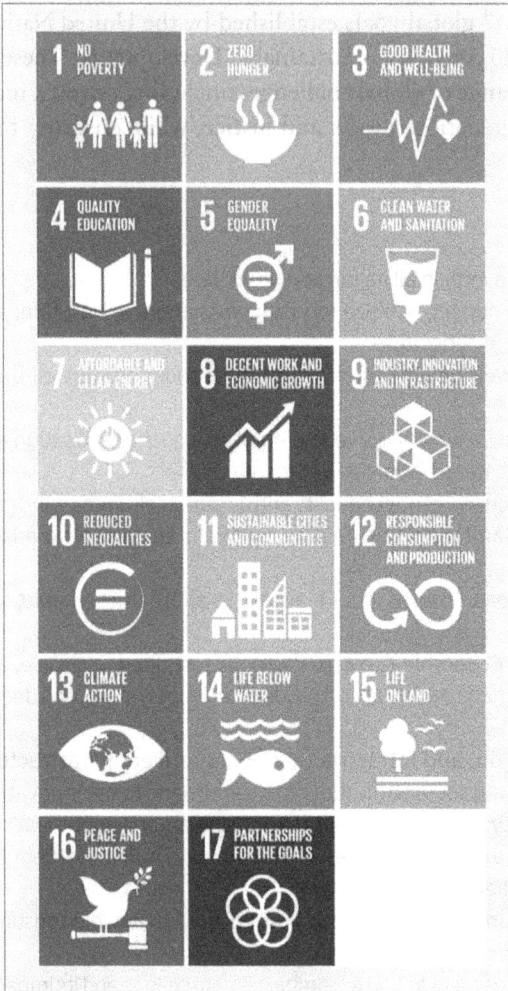

16. **Peace, Justice, and Strong Institutions**: Promote peaceful and inclusive societies for sustainable development, provide access to justice for all, and build effective, accountable, and inclusive institutions at all levels.

17. **Partnerships for the Goals**: Strengthen the means of implementation and revitalize the global partnership for sustainable development.

There have been positive improvements, linked to the implementation of the SDGs worldwide. For instance: global poverty has been significantly reduced since 2000. The percentage of people living in extreme poverty dropped from 15.7 percent in 2010 to 8.6 percent by 2018. However, progress has been uneven, and the COVID-19 pandemic caused a setback. Access to primary education has improved, with enrollment rates rising in many regions, particularly Sub-Saharan Africa. There has also been progress in gender parity in education.

Significant gains have been made in reducing child mortality and improving maternal health. Life expectancy has increased, and more people have access to essential health services. According to the United Nations, countries have adopted various policies to reduce carbon emissions and shift toward green energy. The Paris Agreement, a key step for climate action, is part of these efforts. Efforts have also been made to curb illegal fishing and deforestation.

There is a general consensus however, that the goals will not be achieved by 2030, despite the deployed efforts.

Despite progress, inequality within and among countries remains a significant challenge. The wealth gap between the rich and poor has widened, and millions remain in poverty, especially in low-income countries. The COVID-19 pandemic increased poverty rates and exacerbated inequalities. Climate Change remains an issue: despite efforts, global temperatures continue to rise, and extreme weather events are becoming more frequent. Carbon emissions have not been reduced sufficiently to meet the goals of the Paris Agreement. Political and economic barriers still impede faster progress on climate action. While there have been reductions in hunger in some regions, global hunger has been on the rise in recent years.

As another reality, millions of children, especially in conflict-affected areas, still lack access to schooling. The pandemic also caused significant disruptions to education systems.

Deforestation, habitat loss, and species extinction continue at alarming rates despite international agreements. Efforts to halt biodiversity loss have been insufficient.

Conflicts and violence continue to destabilize many regions especially in the Middle East, in Sahel and in the Horn of Africa but not only. This is undermining the progress in achieving SDG 16 (Peace, Justice, and Strong

Institutions). Human rights violations, corruption, and lack of strong institutions remain significant challenges.

Devastating armed conflicts jeopardize fragile regional stability and hinder progress toward the SDGs.
The conflict in Sudan, which erupted in April 2023 between the Sudanese Armed Forces (SAF) and the Rapid Support Forces (RSF), has plunged the country into a devastating humanitarian and political crisis. Rooted in long-standing power struggles, the fighting has led to widespread violence, mass displacement, and a breakdown of essential services. Civilians have borne the brunt of the conflict, with thousands killed and millions forced to flee their homes, exacerbating an already fragile situation marked by food insecurity and economic collapse. The international community has struggled to mediate a lasting ceasefire, as regional and global actors navigate complex alliances and interests. Meanwhile, the conflict has eroded trust in both national institutions and international diplomatic efforts, raising concerns about Sudan's future stability and the risk of broader regional destabilization. As the war persists, efforts to secure humanitarian access and facilitate negotiations remain critical to preventing further suffering and ensuring a path toward peace and democratic governance.

The conflict in Kivu, located in the eastern Democratic Republic of the Congo (DRC), remains one of the most persistent and complex crises in Africa. Rooted in historical communitarian tensions, competition over natural resources, and the presence of numerous armed groups, the conflict has led to widespread violence, mass displacement, and severe humanitarian suffering. Armed factions such as the March 23 Movement (M23), various militias, and foreign-backed rebel groups continue to destabilize the region, often targeting civilians, engaging in illicit mineral exploitation, and fueling cross-border tensions between the Democratic Republic of Congo DRC and the neighboring countries: Rwanda and Uganda. The Congolese government and international peacekeeping forces struggle to contain the violence and provide security. Despite peace initiatives and diplomatic efforts, Kivu remains volatile, with ongoing human rights abuses, food insecurity, and an escalating humanitarian crisis.

Achieving the SDGs requires substantial investment, which has not been fully met. Low-income countries, in particular, face financial constraints. The global pandemic further stretched resources, diverting funds from SDG-related initiatives.

The SDGs represent an ambitious global agenda to address interconnected challenges which has lost momentum in the current context of fragmentation of the international order. The risks of destabilization amplify poverty which in itself also creates new challenges in terms of health, education, infrastructure, facilities and climate actions. The section below will attempt to provide an analysis about the situation.

II. The Limitations and Barriers: Inconsistent Global Policies, Funding, Accountability, and Lack of Will

The international dialogue on climate change gained momentum in Paris in 2015, when the national governments agreed to reduce climate change by 2°C and possibly 1,5°C compared to the preindustrial area. Since then, there is a common consensus, that the different dialogue platforms have been disappointing in their scope and agreements and as shown by the failure of the COP29 in Baku in 2024. The trajectory of the GHG emissions could lead to an increase of 2°C sooner than expected. Recent projections alert the possibility to reach 6°C of warming in case nothing is done to prevent it. Global surface temperature will continue to increase until at least mid-century under all emissions scenarios considered. Global warming of 1.5°C and 2°C will be exceeded during the twenty-first century unless deep reductions in CO_2 and other GHG emissions occur in the coming decades (IPCC 2021). The reasons for this lack of efficiency in the climate negotiations can be explained by several parameters linked to governance.

A decreased trust in a changing international order

The failure of climate negotiations within the annual Conferences of the Parties (COPS), since the 2015 Paris Agreement seems to reflect a lack of will of national governments to abide by the current international liberal order. According to Féron et al. (2020), there is indeed a dissatisfaction with the current international order from the southern countries (through the angle of inequalities and climate justice), but also from the rising powers (in particular from the BRICs, including the authoritative regimes of Russia and China). In this new untrusted but also competitive multilateral order, it will be increasingly difficult to create approved norms and codes of conduct leading to trusted mechanisms of accountability among national governments (and blocks created by the Western World/China/Russia/Emerging Economies). A new multilateral complex and multipolar world is emerging, but according to the authors, this is not the only current change. "The international world order, whether understood as multiplex, or multipolar, integrates a fragmentation of significant actors such as cities, multinational companies, non-governmental organizations, and different communities enabled by new technological developments" (Féron et al. 2020: xxi, Gaborit 2024). This would be increased by how cross-cultural individual encounters are shaping the cultures of agencies, increasing the fragmentation of an internationalized or globalized environment. This makes it increasingly complex to develop a holistic and global understanding on the needed change to find solutions for the difficult times ahead (of uncertainty, fragmented governance and societies, and climate change).

The resurgence of a self-centered and dominant leadership style exemplified by Donald Trump in the United States, or Vladimir Putin in Russia (with certain limits to the comparison, however), has the potential to reshape geopolitical dynamics, reinforcing a world order rooted in power politics and "rapports de force" (power struggles). Such leadership styles often prioritize national interests and sovereignty over multilateralism, undermining collaborative efforts to address global challenges like climate change. Leaders with unilateralist tendencies may view climate negotiations through the lens of economic competitiveness and geopolitical advantage rather than as a shared responsibility, creating obstacles to consensus. For example, a leader like Trump, who withdrew the United States from the Paris Agreement on the first day of his nomination in January 2025, will inevitably slow or reverse global climate commitments, while Putin's energy-focused policies—reliant on fossil fuel exports—may conflict with decarbonization goals.

Everything appears to be working against the implementation of the SDGs—from the dismantling of USAID to the rollback of funding for UN institutions. Donald Trump's push to revive the fossil fuel industry, combined with his denial of human-driven climate change, threatens to accelerate the pace of environmental disruption. These decisions come with a profound responsibility for the human toll they may exact in the future, potentially setting an irreversible course in motion.

This shift toward a more adversarial global stage could fracture international cooperation, stalling critical progress on climate negotiations and fostering a "zero-sum" approach to climate diplomacy. Ultimately, the resurgence of such leaders may challenge the fragile balance between national interests and the collective action required to address the escalating climate crisis, delaying the ambitious decisions urgently needed for a sustainable future.

Inconsistent or antagonistic policies: Climate initiatives are emerging worldwide and rely on smart technologies and innovations. The construction sector, for example, is developing greener technologies like sustainable concrete, solar panel recycling, and earth materials—all of which offer promising paths toward low-carbon solutions. The emergence of green cement in European countries, with its low-carbon footprint and high recycling potential, represents an encouraging development for urban construction, particularly with its improved resilience to rainfall and heat. However, this optimism faces challenges from antagonistic policies. The negative effects of certain policies on climate adaptation and mitigation—such as deforestation, unsustainable fisheries, and continued investment in fossil fuels and coal mining through subsidies—often counteract the benefits of climate-focused initiatives. This contradiction manifests both internationally, as seen in the challenges of energy transition, and locally, where urban expansion and agricultural demands drive rapid deforestation, the destruction of ecosystems and further land conversions.

The complexity of an approach based on climate risks. Communicating the complex challenges related to climate change presents significant difficulties. According to Moench et al. (2011), policymakers often mistakenly believe that actors will be able to implement solutions once they have been decided. This assumption would imply, for example, that scientists conduct research specifically tailored to local climate conditions. However, in reality, this rarely happens, as research is driven by multiple motivations (financial, academic, scientific, personal interest), factors (knowledge, capacities, access to data), and limitations (available time and tools). Wishful thinking should be avoided in this regard, and it is crucial to acknowledge the vast gap between intentions and implementation.

The responsibility for implementing informed climate adaptation strategies falls primarily on national, regional, and local policymakers, as well as global economic actors. Effective communication and dialogue require tailored and accurate information, which is particularly challenging in the field of climate risk assessment. Climate change and adaptation involve multiple interconnected issues, including water and resource management, as well as the degradation of vital ecosystems such as oceans and forests.

Unfortunately, environmental and climate issues, resource management, economic development, ecosystem protection, and social inclusion often do not rank among political priorities. Similarly, disaster risk prevention, such as wildfire mitigation, is frequently overlooked, despite its devastating impact on communities and ecosystems. Policy implementation generally involves trade-offs, where climate concerns, social inclusion, and resource protection are regularly sacrificed.

This problem can be illustrated by the fable of the King and the Impossible Bridge, which is recalled below.

The King and the Impossible Bridge

Once upon a time, there was a powerful but capricious king who loved to test the intelligence and loyalty of his subjects. One day, he summoned the wisest and most skilled people in his kingdom and commanded:

— "Build me a bridge across the great river, but without stone, wood, or rope. Anyone who fails will be banished from the kingdom."

The architects and engineers looked at each other, bewildered. How could they build a bridge with no materials? Some began to debate, while others sought to flee. But an old sage asked for time before giving his response.

A few days later, he returned to the king and said:

— "Your Majesty, your bridge is ready. But before you cross it, I ask you to show me the shadow of the wind and the sound of silence."

The king stood still, speechless. He realized that he had made an absurd demand, and that the sage had answered with wisdom. Instead of punishing his people for an impossible task, he learned to respect the limits of reason.

From that day on, he became a fairer ruler, and his subjects no longer had to fear unreasonable commands.

But another fable, this time a very real one, is that of a lack of funding and the waste of resources.

The Lack of funding

Let us first address the critical lack of funding on the ground. The reality of development is that, in the face of floods, storms, and even wildfires, communities, cities, and even entire states struggle with a severe shortage of resources.

We witnessed this in Mayotte, when Cyclone Chido devastated the island in late 2024; in Florida, where Hurricanes Helen and Milton caused significant human and material losses; and in Valencia, Spain, after the deadly floods of November 2024.

While funding theoretically exists, the real question is: where is it, to what extent is it available, and—most importantly—who actually has access to it?

Cyclone Chido struck Mayotte on December 14, 2024, with wind speeds exceeding 200 km/h and gusts over 225 km/h, making it the strongest storm to hit the island in at least 90 years. The confirmed death toll is 39, but the actual number may be higher due to challenges in accounting for undocumented migrants

The funding that the countries will need to invest in dealing with climate change is estimated to be around 9 percent of its GDP on infrastructure development during the period of 2016–2030.[70] At the recent COP in Baku, Azerbaijan the United Nations came up with the estimate that 1,000 billion a year would be necessary in investments. The approved agreement proposed that developed nations provide $300 billion annually by 2035 to assist developing countries in addressing climate change. This figure is part of a broader goal to mobilize $1.3 trillion per year from different sources.[71]

However, this proposal has been met with criticism from developing nations, who argue that the amount falls short of their needs. For instance, China and the G77 group of developing countries have called for at least $500 billion annually.[72]

The disparity between these figures highlights ongoing tensions and the complexity of reaching a consensus on climate finance at COP29.

---

[70] Chapter 4 Infrastructure Investment and the Sustainable Development Goals

[71] <https://www.politico.com/news/2024/11/22/climate-proposal-would-see-rich-countries-pay-250-billion-a-year-00191216?utm_source=chatgpt.com>

[72] <https://www.ft.com/content/3b4299fc-84cd-4f48-90d1-bbd1a811e07e?utm_source=chatgpt.com>

But the situation may lead to risky trade-offs, focusing on what the public governments can truly tackle and implement (e.g., green nature-based solutions, domestic waste) to the detriment of more long-term climate adaptation policies: to protect populations that are the most exposed to hazards (protection of forests and oceans, agriculture resilience programs, protection of transport and networks, contingency plans, social programs etc.).

There are also justified criticisms that the funding can be wasted in inefficiencies, corrupted systems, lack of coordination, prioritized payment of elites over investments for communities, and in administrations lacking competence.

Local climate adaptation strategies depend on **limited available local funding.**

The question of funding is also linked to the question of bureaucracy that is jeopardizing the necessary changes to implement climate efficient policies.

Bureaucracy and the waste of funding

The fable below is the one of modern times that many international experts including myself will recognize as one of their experiences.

The King's Golden Well

Once upon a time, in the Kingdom of Solara, a terrible drought struck. The rivers dried up, the crops withered, and the people cried out for water. Moved by their suffering, the king summoned his advisors.

— "We must act! Let us build a grand well to provide water for all!"

The royal court burst into applause. Gold from the kingdom's treasury was poured into the project. Architects, consultants, and foreign experts were brought in. Surveys and studies were commissioned, grand reports were written, and lavish banquets were held to discuss "best practices."

Years passed. The golden well project became famous. Dignitaries from distant lands came to admire its plans. But when the villagers went to draw water, they found… nothing.

— "Where is the well?" they asked.

The officials shuffled uncomfortably.

— "Well, the initial funds were spent on project assessments. Then, part of the budget went to administrative costs. Of course, we needed international oversight, so we hired specialists. And we built a magnificent structure, but unfortunately, we ran out of money before reaching the water table."

The king, ashamed, finally visited the site. He saw a grand marble arch, a shiny plaque listing the donors, and a dry hole in the ground.

The villagers, still thirsty, had already begun digging their own small wells with simple tools.

Moral of the story: A project is only as valuable as its results. When resources are spent on appearances rather than impact, those in need are left with nothing but promises.

Accessible funding for developmental and humanitarian projects, as well as for research and development (R&D) are often significantly hindered by excessive bureaucracy, convoluted reporting mechanisms and numerous conditions. This can lead to delays in the implementation of projects, and even a complete stagnation of crucial initiatives. The labyrinth of administrative procedures, stringent documentation requirements, and slow approval processes can sap resources that are meant for immediate relief or long-term development. These bureaucratic hurdles are not only frustrating for those on the ground trying to implement projects but also jeopardize the sustainability of actions and programs, preventing them from achieving their intended outcomes.

International aid (e.g., for the European Commission) has been characterized as a prototypical example of red tape. The stringent rules around grant applications, reporting obligations, and financial accountability are often so cumbersome that they can delay critical projects or dissuade potential applicants from pursuing available funding altogether. NGOs, research organizations, the private sector and other beneficiaries frequently struggle to meet the high bureaucratic standards required, which diverts focus and resources away from their core missions, often forcing them to allocate more time and staff toward administrative tasks rather than direct action as also illustrated by the 2024 Draghi report.[73]

However, the problem of excessive bureaucracy is not unique to specific donors. Similar challenges are reported with funding from other major

---

[73] <https://commission.europa.eu/topics/strengthening-european-competitiveness/eu-competitiveness-looking-ahead_en>

international agencies, such as the United Nations. World Bank-funded projects, which play a pivotal role in global development, also face bureaucratic obstacles (World Bank, 2005).

This case is emblematic of a broader issue: during crises, such as natural disasters or humanitarian emergencies, time is of the essence, yet bureaucratic processes can significantly slow down recovery efforts. These delays have real-world consequences, impeding not only short-term recovery but also the long-term sustainability of initiatives. Furthermore, the need to navigate various layers of bureaucracy can often lead to inefficiencies, duplication of efforts, and miscommunication between stakeholders.

This bureaucratic entanglement can discourage smaller organizations or local grassroots initiatives from applying for international funding, as they often lack the institutional capacity to navigate complex administrative requirements. This creates a paradox where those most in need of funding, particularly in low-income or disaster-stricken regions, may find themselves unable to access it due to the very systems designed to help them. Thus, the very mechanisms intended to ensure transparency, accountability, and proper use of funds sometimes become barriers to effective action. In conclusion, while bureaucracy is essential for maintaining accountability and ensuring that funds are properly allocated, the current systems often obstruct timely and sustainable interventions. Efforts to streamline administrative processes and provide more flexible, accessible pathways to funding are needed to prevent bureaucratic bottlenecks from undermining the potential for impactful and lasting change.

Accountability: in terms of climate adaptation and climate mitigation, the increasing identified problem will soon become the question of accountability.

In the area of climate adaptation, the question of accountability is becoming increasingly critical. The floods in Liège Province in July 2021 and in Valencia in October 2024 demonstrate how difficult it is to identify responsibilities after climate-related disasters. For instance, the parliamentary inquiry following the Liège floods revealed that no single actor or stakeholder could be held responsible for the disaster and its impacts. Instead, the inquiry identified a chain of dysfunctions and cascading failures: first responders lacked proper equipment, city mayors waited for provincial instructions before starting evacuations, and dam authorities failed to fully optimize flood control. This situation not only left questions about responsibility during the unprecedented flood but also created confusion about accountability for recovery and reconstruction, with some victims still awaiting compensation or financial support. This example highlights the need for comprehensive accountability systems where all government tiers, public and private actors are held responsible for their actions, policies, and decisions.

III. Toward the Emergence of Strong Trust Community Networks to Implement SDGs and Climate Adaptation

Addressing climate adaptation through governance presents significant challenges. Effective governance requires more than just new forums for dialogue—it demands consistent policy-making, robust accountability frameworks, and inclusive engagement of diverse stakeholders. While authoritarian regimes claim greater effectiveness in implementing policies through swift, unconstrained action, evidence proves otherwise. Analyses of climate actions at national and city levels under authoritarian regimes show poor results. Their growing environmental footprint, compared to the better emission reduction achievements of democratic systems, reveals their ineffectiveness.

This failure stems from the rigid hierarchical structures of authoritarian regimes, which restrict bottom-up solutions. Elite decision-making, constrained by regime loyalty, stifles creativity and blocks innovative grassroots initiatives. As Charles Tilly observes, "authoritative organizations often produce few collective benefits, impose large collective costs, and primarily use resources to perpetuate themselves, regardless of public suffering or discontent" (Tilly, 2005:40). Though hierarchical organizations can generate benefits through top-down incentives like coercion or capital, they fail to foster the collaborative approaches needed for effective climate governance.

In contrast, collaborative institutions—based on mutual consent among members—offer an inclusive and flexible approach. These institutions, from international organizations to local resource management initiatives, operate through cooperatively developed rules. As Ostrom (1998) notes, lasting collective cooperation depends on three reinforcing factors: interpersonal trust, investment in reputation, and reciprocal norms. These elements strengthen accountability by fostering trust, respect, and adherence to law.

At the local level, "trust networks" are vital for climate action. These networks, defined as "bounded, internally communicating sets of relations entailing mutual obligations" (Tilly, 2005:44), build on strong social ties. Environmental NGOs and transnational networks demonstrate how trust-based systems drive positive change. The partnership between collaborative institutions—local governments, climate agencies, and civil society—and trust networks enables effective local climate policies.

Participatory approaches and civic engagement play crucial roles. Research shows that cooperation among diverse stakeholders—citizens' associations, nonprofits, and forums—drives climate adaptation efforts (Hegger et al., 2017; Marchezini, 2020). The Netherlands offers examples where local governments have engaged residents in stormwater management and heat stress prevention. Yet gaps remain, including weak citizen engagement in flood preparedness and insufficient solidarity networks for proactive measures (Hamman, 2011).

The COVID-19 pandemic exposed the weakness of solidarity networks. Many proved inadequate for supporting vulnerable groups, especially the elderly and youth, while existing networks weakened under lockdowns. After recent floods, post-disaster solidarity networks emerged for food distribution but remained fragmented and poorly coordinated. Problems like price gouging and exploitative insurance practices highlighted the need for resilience measures combining institutional preparedness with strong community systems.

Finally, limited private sector and financial institution involvement in clean energy solutions reveals another trust gap. This mistrust between governments, regulators, and stakeholders hinders cooperation and slows sustainable innovation. Building trust networks to bridge these divides is essential for fostering collaboration, reducing risks, and ensuring effective climate action.

Addressing climate change through local and national efforts depends on cooperation, strong social networks, and trust among stakeholders. Yet these crucial foundations are weakening. While multi-level governance systems exist to advance climate adaptation and mitigation, their practical effectiveness falls short of their promise.

Corporate and government greenwashing undermines public confidence through empty "green" initiatives that mask inaction. Social media amplifies

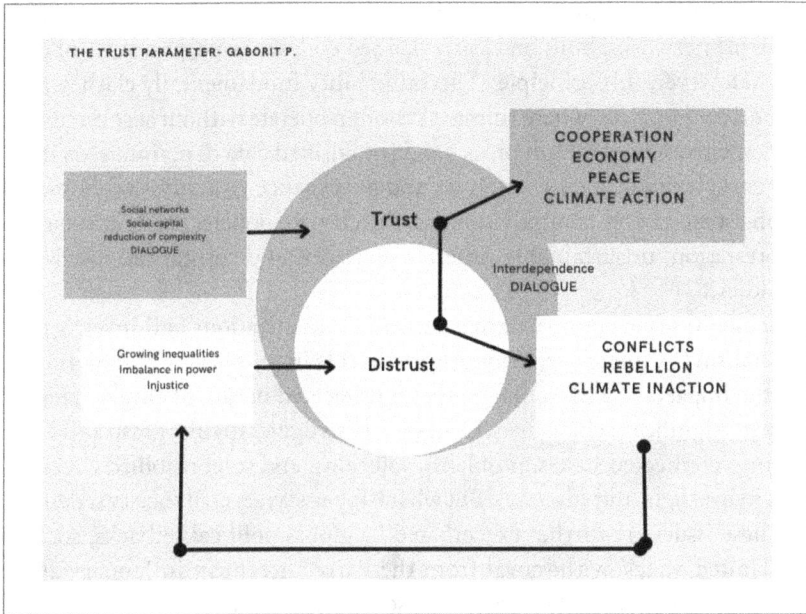

Figure 5. The trust parameter. Author Gaborit Pascaline

misinformation, while targeted disinformation campaigns create confusion, erode trust, and divide public opinion. Rather than building shared urgency, these forces create division and block real progress. Many organizations, driven by short-term profits or political advantage, focus on appearances while delaying crucial action on the worsening climate crisis.

This situation exposes a stark truth: the essential elements of collective climate action—trust, cooperation, and accountability—are giving way to opportunism and deception. Without rebuilding these foundations and fighting disinformation, our chances for meaningful climate progress diminish daily, leaving us vulnerable to devastating consequences.

Conclusion

Climate adaptation and mitigation at national and local levels, as well as the successful implementation of the SDGs, hinge on robust accountability systems. These systems require the establishment of collaborative institutions and trust networks, built on clearly defined codes, norms, and rules of conduct. However, this principle of accountability fundamentally clashes with the current practices, where ruling elites often operate without accountability to their citizens and rely on propaganda or falsified data to maintain control. The complexity of their economies and governance structures often masks inconsistencies and contradictions in the climate policies, such as ongoing deforestation, unsustainable fisheries practices, and substantial fossil fuel subsidies.

While in some countries environmental organizations and investigative journalism fortunately still can play a vital role in uncovering and highlighting the impacts of these policies, the multifaceted nature of climate change presents significant challenges. Civil society struggles to fully grasp and track the interconnected stakes, problems, solutions, and responsibilities, leaving gaps in oversight and advocacy. But what happens when civil society is denied?

These issues are further exacerbated by global political setbacks, such as the United States' withdrawal from the Paris Agreement in January 2025. This decision weakened international solidarity and undermined the shared commitments critical to combating climate change, sending mixed signals about the urgency of global climate action. Moreover, scientific uncertainties surrounding the best solutions for climate adaptation and mitigation also hinder trust between stakeholders. Conflicting scientific recommendations, debates over the effectiveness of proposed strategies, and the lack of consensus on priorities often leave policymakers and the public divided, further complicating efforts to build cohesive and actionable frameworks.

In an increasingly polarized global system, marked by intensifying tensions and growing distrust between countries and nations, these factors exacerbate divisions and erode the foundations of international cooperation. Ultimately, such fractures hinder collective progress and disproportionately harm the countries and regions most vulnerable to the devastating impacts of climate change. Bridging these divides and fostering genuine collaboration will require a renewed commitment to transparency, mutual trust, and shared responsibility—values that remain fragile but essential for the future of global climate governance.

But Imagine a world ravaged by storms, floods, and famines, where humanity has destroyed its greatest ally—the Earth.

References

Afrizal, Berenschot W. et al. 2020 "Resolving Land Conflicts in Indonesia" Review essay, Bijdragen tot de taal, land en volkenkunde 176 (2020) pp. 561–574.

Basset T., Fogelman C. 2013 "Déjà vu or something new? The adaptation concept in the climate change literature," *Geoforum* 48, 42–53.

Braithwaite V., Levi M. 1998 *Trust and Governance*, Russell Sage Foundation series on Trust.

De Arauso Arosa Monteiro R., Ferraz de Toledo R., Roberti Jacobi R. 2021 "Dialogue Method: a proposal to Foster intra and inter-community Dialogic Engagement," in *Journal of Dialogue Studies*, Special issue, Dialogue with and among the Existing, Transforming and Emerging Communities, Vol. 9, 165–183.

Djalante R., Garschagen M., Thomalla F., Shaw R. (eds.) 2017 *Disaster Risk Reduction in Indonesia*, Springer International publishing.

Féron E., Juutinen M., Käkönen J, Maïche K. (eds.) 2020 "Shedding light on a Changing International Order: Theoretical and Empirical Challenges", Tampere TAPRI Press.

Funtowicz S. 2020 "From Risk calculations to narratives of danger," in *Climate Risk Management*, Vol. 27 *100212* (2020), <https://reader.elsevier.com/reader/sd/pii/S2212096320300024?token=B6CBEA02054999EE756B04F666D4701C3905CDF460510AD1F52DF26A57DFF692EC212672A58C4D7A49A536076F984FAB&originRegion=eu-west-1&originCreation=20210808171814>, last accessed 08.08.2021.

Gaborit P., Aleksic A., Marengo P., Diab Y., Pathak K. 2020 Policy briefs ten Pilot Cities <https://www.resilient-cities.com/en/knowledge/175-policy-briefs-for-pilot-cities-2> last accessed 31.03.2021.

Gaborit P. 2021 "Vulnerabilities and Resilience to Climate Change in Tanzania" in Gaborit P. et Olomi D. (eds.) *Learning from resilience strategies in Tanzania: an outlook of international development challenges*, Brussels, Peter Lang International, <https://www.peterlang.com/document/1152350>

Hamman P., Causer J. Y. 2011 *Villes, environnement et transactions démocratiques* (Dir.) Peter Lang, Ecopolis.

Hardin R. 2004 *Distrust*, NYC, Russell Sage Foundation.

Hegger D. L. T., Mees H. L. P., Driessen P. J., Runhaar A. C. 2017 "The roles of residents in Climate Adaptation: A systematic Review in the case of the Netherlands," in Environmental Policy and Governance <https://doi.org/10.1002/eet.1766>

IPCC international Panel on Climate Change, 2021, 6th assessment report: Climate Change the Physical Assessment report, <https://www.ipcc.ch/report/ar6/wg1/downloads/report/IPCC_AR6_WGI_SPM_final.pdf>

Koch L., Gorris P., Pahl Wostl C. 2021 "Narratives, Narration and Social Structure in environmental Governance," in *Global Environmental Change*, Vol. 69, July 2021, 102317, <https://doi.org/10.1016/j.gloenvcha.2021.102317>, last accessed 23.07.2021.

Luhmann, N. 1979 *Trust and Power: Two Works by Niklas Luhmann. Translation of German originals Vertrauen 1968 and Macht 1975.* Chichester: John Wiley.

Marchezini V. 2020 "What is a sociologist doing here?: an unconventional People centered approach to improve warning implementation in the Sendai Framework for Disaster Risk Reduction" in *Int J Disaster Risk Sci* 11:218–229, <https://doi.org/10.1007/s13753-020-00262-1> last accessed February 12, 2022.

Misztal B. A. 1992 "The Notion of Trust in Social Theory", *Policy, Organisation and Society*, 5:1, 6–15, <https://doi.org/10.1080/10349952.1992.11876774>

Moench M., Tyler S. Lage J. 2011 "Catalyzing urban governance: applying resilience concepts to planning practice in the ACCCRN Program 2009–2011," ACCRN publication.

Mulkhan U., Mayaguezz H., Tisnanta H. S., Kurniawan N. 2020 "Urban Analysis Report Pangkal Pinang" <https://www.resilient-cities.com/en/?preview=1&option=com_dropfiles&format=&task=frontfile.download&catid=41&id=40&Itemid=1000000000000>, last accessed 31.03.2021.

Ostrom E. 1990 *Governing the commons: the evolution of institutions for collective action*, Cambridge University Press.

Putnam R. 1993 *Making Democracy work: Civic traditions in Modern Italy*, Princeton University Press.

Rosenzweig C., Solecki W. D., Romero-Lankao P., Mehrotra S., Dhakal S., and Ali Ibrahim S. (eds.) *Climate Change and Cities: Second Assessment*

Report of the Urban Climate Change Research Network, Cambridge University Press New York.

Szescynski B. 1999 "Risk and Trust: The Performative Dimension," in *Environmental Values* pp. 239–252 The White Horse Press Cambridge UK.

Tilly C. 2005, *Trust and Rule*, Cambridge University Press.

Wijaya N., Nitivattanon V., Prasad Shrestha R. and Minsun Kim S. 2020 "Drivers and Benefits of Integrating Climate Adaptation Measures into Urban Development: Experience from Coastal Cities of Indonesia", *Sustainability* 2020, 12, 750, Mdpi.com.

World Bank 2005 "Rebuilding a better Aceh and Nias: stocktaking of the reconstruction effort," world bank.

Ziervogel G., Pelling M., Cartwright AA; Chu E., Deshpande T, Harris L … Zweig P. 2017 "Inserting rights and justice into urban resilience: a focus on everyday risk," *Environment and Urbanization*, 29(1), 123–138.

Trust Barriers Amid Global Tensions

Imagine a world divided by trade conflicts where only the interests of a few would prevail—and where tensions could escalate into conflict ...

Trust is essential—not only for initiating action but also for implementing effective, inclusive, and sustainable policies. It serves as the foundation for collaboration, enabling diverse stakeholders to work together toward shared goals, particularly in the face of complex and urgent global challenges such as climate change, public health crises, and geopolitical tensions. As discussed in previous chapters, trust plays a pivotal role in crisis resolution, whether during unforeseen social conflicts or in situations characterized by heightened tensions, misinformation, and escalating narratives. Trust is not merely an abstract or theoretical concept—it is a practical necessity that underpins dialogue, decision-making, and collective action. Without trust, even the most well-designed policies are likely to face resistance or fail entirely in their implementation.

However, we are currently navigating a geopolitical phase marked by the systematic deconstruction of trust. This "trust deficit" manifests in numerous ways: the erosion of faith in institutions, growing skepticism toward governments and international organizations, and the rise of polarized narratives fueled by misinformation and disinformation campaigns. Geopolitical rivalries, ideological divides, and a lack of transparency further deepen this crisis of trust, making it increasingly difficult to foster the cooperation and accountability required for addressing global challenges. Rebuilding trust in this fractured landscape is not only a moral imperative but a strategic necessity to ensure effective governance, meaningful progress, and a more stable and resilient global order.

* * * *

In an increasingly fragmented or fractured geopolitical order dominated by power struggles, negotiations are becoming both a necessity and a gamble. During crises, actors may be forced into negotiations out of sheer necessity, yet this compulsion does not guarantee success or good faith. Conflict

protagonists—driven by opposing interests over resources, trade, or influence—may, at certain moments, share a reluctant interest in cooperating.
However, such moments often emerge only when the situation reaches a stale-
mate, and no party sees a clear path to victory. While crises may sometimes
create conditions conducive to negotiation, the balance of power between
parties and their genuine interest in resolving the conflict are critical precondi-
tions (Thuderoz, 2003). Yet, even when these conditions are met, negotiations
should never assume that all parties approach the table in good faith or with
equal trustworthiness (Stedman, 2003).

Negotiations inherently require trust (Thuderoz, 2003), but trust is
increasingly rare in a world where actors routinely prioritize self-interest
and exploitation over cooperation. Trust involves significant risk—negotiators must gamble not only on the opposing party's willingness to honor
agreements but also on their ability to maintain credibility with the groups
they represent. As Niklas Luhmann (1979) posited, trust is fundamentally
tied to risk-taking, yet in the modern geopolitical landscape, the risks of
betrayal, deception, or exploitation often outweigh the potential rewards
of cooperation. This is particularly evident in social conflicts, where local
communities frequently feel betrayed by negotiators who prioritize private
or external interests over the well-being of their constituents.

The negotiation process itself is fragile and vulnerable to sabotage.
Dishonest intentions, manipulation, and external interference can easily
disrupt trust-building efforts. The "spoiler" theory (Stedman, 1997; Greenhill
et al., 2007) highlights how individuals or groups with vested interests in
maintaining conflict will deliberately derail negotiations. These spoilers,
perceiving negotiations as threats to their power, worldview, or control,
actively seek to undermine dialogue. Additionally, parties may resort to
finding scapegoats—identifying external minorities or powerless groups to
blame for the conflict—as a strategy to deflect responsibility and create a
temporary illusion of unity (Deutsch, 1958). While this tactic may strengthen
short-term trust among negotiators, it perpetuates long-term divisions and
reinforces systemic injustices.

The role of third-party mediators and institutions in resolving crises is
similarly fraught with challenges. Effective mediation requires trust in the
institutions overseeing the negotiations. This trust depends on the institution's
perceived neutrality, legitimacy, and shared adherence to rules, values, and

standards by all parties. Yet, in an era of growing distrust in institutions—fueled by inconsistent actions, biases, and external pressures—such trust is increasingly hard to establish. Institutions themselves are often seen as extensions of power struggles, rather than as impartial facilitators of peace or compromise.

When conflicts involve multiple actors or complex stakes, whether they are geopolitical, trade-related, or social in nature, dialogue becomes a critical yet fragile tool. While dialogue may help foster a more trustworthy environment, it is often insufficient in resolving crises, especially in a world where trust is continuously undermined by manipulation, disinformation, and competing interests. Without genuine commitment to cooperation and accountability, negotiations risk becoming another arena for power plays, further entrenching divisions and prolonging conflicts rather than resolving them.

Trade Conflicts

Trade conflicts arise when nations impose restrictive policies or tariffs that disrupt the free flow of goods and services, often as a response to perceived unfair practices or economic imbalances. These disputes can stem from accusations of currency manipulation, IP theft, or subsidies that create an uneven playing field in global markets. While some countries see protectionist measures as a way to shield domestic industries and preserve jobs (and answer the Wheeler Security Dilemma), such actions often provoke retaliatory measures, leading to escalating tensions. Trade wars can undermine global economic stability, disrupt supply chains, and increase costs for businesses and consumers alike.

Resolving these conflicts requires diplomatic negotiation, adherence to international trade rules, and a willingness to find mutually beneficial solutions. In an interconnected world, cooperation often proves more valuable than competition in achieving long-term prosperity.

The trade war between the United States and China spilling out to other countries including Europe, has become a defining feature of modern global economic relations, driven by conflicting interests, strategic rivalries, and shifting power dynamics. The United States has long accused China of engaging in unfair trade practices, including IP theft, state subsidies, and restrictive market policies. This has led to the imposition of tariffs on hundreds of billions of dollars' worth of Chinese goods, to which China has responded with retaliatory tariffs, escalating the standoff. Meanwhile, Europe has found itself entangled in its own trade disputes with the United States, particularly in industries such as aerospace, steel, and agriculture, rooted in disagreements over subsidies, trade imbalances, and market access. Although Europe and China have occasionally aligned in resisting U.S. protectionist policies, their own frictions over issues like technology transfers, human rights concerns, and market openness add another layer of complexity to this trilateral rivalry.

Under Donald Trump's renewed leadership, U.S. trade policy has taken an even more unpredictable turn, creating uncertainty on a global scale. Trump has revived his "America First" approach, advocating for increased tariffs, reshoring of industries, and reducing dependency on China, while threatening punitive measures against European industries if they fail to align with U.S. interests. These policies have disrupted traditional alliances, making it difficult for trading partners to rely on stable agreements. Trump's willingness to impose sudden tariffs, withdraw from multilateral trade deals, and use trade as leverage in geopolitical disputes further exacerbates global economic tensions. This unpredictability undermines trust in the United States as a reliable economic partner and forces other countries to hedge their strategies, accelerating decoupling trends and fragmenting global trade.

On February 1, 2025, President Donald Trump announced the imposition of new tariffs aimed at combating illegal immigration, drug trafficking, and trade imbalances—only to later suspend the decision. On February 13, 2025, the U.S. government announced its intention to implement a reciprocal tax regime on imports. On April 2, 2025, the U.S. president unexpectedly revealed a detailed plan for reciprocal tariff measures—with surcharges ranging from 10 percent to 50 percent—calculated not from actual tariffs but based on regulatory barriers and market protection measures. These tariffs, if implemented, would have severely impacted the global economy and emerging countries trying to protect their industries. Though the U.S. president later withdrew measures targeting most countries due to financial market reactions, China and the United States still entered a tariff escalation. The Trump administration imposed a 145 percent surcharge on Chinese products entering the United States, though later exempted certain electronic products. China retaliated with a 125 percent surcharge on U.S. imports. Both countries later came to an agreement, and the average US tariffs on Chinese exports stands in September 2025 at 57.6% and cover all goods, while China's average tariffs on US exports is 32.6% also covering all goods. But the saga continues ...

The new tariffs, symptoms of an emerging trade war, are expected to raise prices for U.S. consumers. Products such as cars, food, alcohol, and electronics will be affected, leaving households to face consequent additional costs annually... Several countries are seeking diplomatic solutions, while others have announced proportional responses. Canadian officials, for instance, did not hesitate to announce retaliatory tariffs on U.S. goods. China has responded with prohibitive tariffs on U.S. imports into its market, India strengthened its ties with other countries including Russia and China but still faces 50% of tarifs on its exports to the U.S., while the EU faced with 15% tarifs could adopt a diplomatic negotiating approach.. Economists warn that these actions could disrupt global supply chains, elevate inflation, and strain relationships with key trading partners.[74] Moreover, this situation jeopardizes directly the rules of the World Trade Organization.

Another chapter in these tensions had emerged with the EU's decision to impose tariffs on China's electric vehicles, accusing Beijing of providing unfair subsidies to dominate the market. While this move reflects Europe's growing efforts to protect its industrial base, some critics argue that the EU has been slow to act in safeguarding its economic interests and investing in ethical research and development to compete on a global scale. This hesitation has left European industries vulnerable to foreign competition and global shifts in trade dynamics. Cooperation with the rest of the world becomes crucial.[75] The Competitiveness Compass, introduced by the European Commission on January 29, 2025, is a strategic framework aimed at boosting the EU's economic strength and ensuring sustainable prosperity. Based on the conclusions of the Draghi report, it sets key priorities for 2024–2029, including simplifying regulations, promoting innovation in green tech and AI, enhancing infrastructure, and supporting European industries through investment and industrial policy. The initiative proposes an annual €800 billion investment to drive productivity and the ecological transition. However, it faces political and legal challenges, including the need for consensus among member states.

---

[74] <https://apnews.com/article/trump-tariffs-trade-china-mexico-canada-inflation-753a 09d56cd318f2eb1d2efe3c43b7d4>, <https://nypost.com/2025/02/01/us-news/trump-begins-long-await-tariffs-on-canada-mexico-and-china/?utm_source=chatgpt.com>

[75] The Global Gateway notably is the EU's international investment strategy, launched in December 2021, to enhance global infrastructure, economic development, and connectivity. It aims to mobilize up to €300 billion by 2027 to support sustainable projects in digital infrastructure, climate and energy, transport, health, and education.

These revolutions in trade policies jeopardize the current global trade rules, highlighting a broader trend where major economies are testing the World Trade Organization boundaries to protect domestic industries.

These escalating trade tensions have had widespread consequences. Global supply chains have been disrupted, consumer costs have risen, and international alliances have been reshaped, with trade policies increasingly becoming tools of geopolitical strategy. The polarization of economic models—China's state-driven capitalism, the U.S.'s market-driven protectionism, and Europe's regulated liberalism—has made multilateral negotiations fraught with challenges, leaving little room for compromise.

Societies and countries are further exposed to growing tensions, centrifugal forces, identity-based divisions, and conflicting interests. While the past 70 years were marked by relative global cooperation and progress, the current trajectory raises concerns about the prospects for peace, stability, and sustainability. In such a fragile and polarized global environment, Trump's unpredictable policies and their ripple effects add another layer of complexity, making trust an even more critical variable. Understanding its role in economic and geopolitical relationships is essential for analyzing the challenges ahead and identifying pathways to foster cooperation, stability, and sustainable development.

Can Dialogue Still Be an Option?

Unlike negotiations, which often take place behind closed doors, dialogue can involve a broader range of stakeholders.. Dialogue here refers to creating spaces for institutional exchanges among stakeholders within a problem-solving framework, leading to "different ways of thinking, talking, learning, and acting" (de Araujo et al., 2021). It involves learning together and identifying solutions and obstacles in a problem-solving manner, while also encouraging knowledge exchange. This dialogue can focus on environmental issues, specifically water management, acknowledging needs and recognizing possible disagreements on values, ideas, or interests. It can also incorporate different narratives, resistances, antagonisms, and demands.

Many authors emphasize the importance of dialogue in conflict resolution especially related to trade wars. Constructive change often emerges through cooperation and social interaction between involved parties. However, in cases of armed conflicts, high tensions or global threats, dialogue may become impossible.

There is also a consensus that solving global challenges and problems require consistent cooperation among all stakeholders, including the private sector in the case of trade, but also the communication channels toward civil society. Dialogue forums are necessary, while at the same time international negotiations have been going backward since the COVID-19 crisis, when the populations and countries have realized the fragility of the supply chains, and the necessity to protect their own interests. Optimism has shifted away, and in a way, optimism was also in itself creating problems as optimists tended to believe that we could achieve SDGs, peace, prosperity and climate goals without implementing any changes. Coherent responses to important problems require involving a multitude of stakeholders in transformative processes, leading to the development of policy and action pathways. However, this cooperation faces many obstacles, with trust being just one factor.

One major challenge is that several crises linked to the management of resources or trade negotiations are crossing geographical territories, communities, and different government jurisdictions. For instance, trade tariffs for Europe are decided by the European Commission, which decreases national governments' legitimacy in the eyes of their populations. At the same time, tax and social systems remain entirely national, undermining trust in Europe's ability to implement fair policies.

Another obstacle is that different organizations have varying interests, goals, and visions which hinder the emergence of solutions for sustainability. This issue has been highlighted by various authors (Burch et al. 2014, Van Bruggen et al. 2019, Harris et al. 2017). It seems impossible to achieve a country's objectives at all scales, communities, and sectors simultaneously, such as water, energy, food, carbon, stable institutions, and peaceful societies.

Explaining the Barriers to Trust

If we have seen earlier that trust is necessary to undermine or exit a crisis, it is a complex matter to facilitate and to reach trust conducive environments which would enable the exit of the crisis. The barriers to trust involving fear, suspicion or distrust can be important parameters to consider in managing a crisis.

Schizophrenic dialogue is a major problem both in trade and international negotiations including over climate. The concept of a schizophrenic dialogue between countries reflects the contradictory and fragmented nature of modern international relations, where states simultaneously engage in cooperation and conflict. Nations may negotiate climate agreements while pursuing policies that undermine environmental goals, or advocate for free trade while erecting protectionist barriers. This duality often arises from competing domestic and international pressures, where governments must balance populist demands, economic interests, and geopolitical ambitions. For instance, countries may pledge allegiance to multilateralism in public forums while pursuing unilateral actions behind closed doors. Such inconsistencies erode trust, complicated diplomatic negotiations, and stall collective action on global challenges. To overcome this fragmented discourse, nations must align their rhetoric with their policies, fostering transparency, mutual understanding, and a commitment to long-term solutions rather than short-term gains.

Trade-Offs and Short-Term Views: A trade-off involves balancing between one or more desirable but sometimes conflicting plans, policies, or measures (Grafakos et al. 2019, Gaborit 2022a). Our field research shows that policies and actions are influenced by choices and priorities, leading to trade-offs and the exclusion of certain sectors or priorities (Gaborit 2022). At the international level security issues will also prevail, which is understandable. What is less understandable is why short-term economic interests benefiting only a few take priority over SMEs, social welfare, health, prosperity, innovation, and sustainable development. Short-term views often get the preponderance over long-term views in that area. Elites are also maintained to power by populations which are deeply affected by economic crisis and inflation, and by a powerful private sector led by multinationals. They cannot therefore prioritize goals differently in the short term...

Security: The security dilemma (Wheeler 2011), is a core concept in international relations, highlighting how actions taken by a state to ensure its own security can inadvertently threaten other states, leading to a cycle of tension and escalation. For example, when one country builds up its military capabilities to defend itself, neighboring states may perceive this as a potential threat and respond by bolstering their own defenses. This mutual mistrust often results in an arms race, where efforts to achieve greater security paradoxically make all parties less secure. The dilemma is particularly acute in regions with historical rivalries or in the face of emerging technologies like cyber warfare and AI, where offensive and defensive capabilities are difficult to distinguish. Addressing the security dilemma requires building trust through transparency, confidence-building measures, and cooperative frameworks that allow states to pursue security without undermining the stability of others.

"Us versus Them": As we have seen in the case of the Baku COP conference, the increasing polarization of the world community between "us" and "them" is lacking a constructive approach. Indeed, the reduction of the problems to a simple question of "money transfer" between "rich" and "poor" countries, the "global north" and the "global south" can lead to increasing tensions in the future, and be used as a manipulative strategy to undermine efforts.

Communication and Disinformation: The creation of trust-conducive environments requires consistent communication and reliability in meeting the different challenges, as well as in fulfilling expectations. Despite challenges in effectively reaching diverse communities and stakeholders, the lack of a consistent engagement process and failure to include community grassroots initiatives in local plans can lead to high costs and consequences. For instance, it can result in technically or socially unsuitable solutions. In several authoritarian regimes, however, disinformation has become the norm, creating a real distance between people and power, the former accepting that lies are part of the government's normal action. The questions of information and legitimacy are, however, important, as detailed in our chapter on disinformation (Chapter 3). Harris (2017) notes "As research and negotiation are conducted behind closed doors, the general public's confidence in scientific, technical and administrative expertise is destined to be low. Without more inclusive processes and lasting mechanisms of social learning and public involvement, even scientific findings, however accurate, fail to gain social legitimacy" and will be caught up in the waves of disinformation. Local actions need to align better with local realities, values, and norms, achievable only through solid local multi-stakeholders' engagement.

Engagement with the private sector but also with communities is crucial especially in the sectors of trade and the environment. The private sector is often involved in pollution and deforestation, (e.g., plastic production, use of energy, waste dumping, water and air pollution), but also leads to innovation. Effective communication and awareness-raising are crucial, treating beneficiaries as active participants in dialogue rather than passive targets of stakeholder-driven messages. Some authors highlight pitfalls in communication, which might simplify situations and spread incorrect messages, serving some stakeholders' interests over others (Brulle 2010,). Engagement and awareness therefore require simple, clear, and comprehensive resources accessible to the public. The absence of messages at all can lead to a lack of risk awareness, especially in vulnerable groups, or justify denial of climate change and climate impacts. It is crucial to engage vulnerable people in decision-making for social justice.

Achieving clarity is challenging in complex sectors, such as in conflicts' resolution, prevention and in climate action where uncertainties are inherent. The different stakeholders often lack clear steps for contributing to solutions. When they do contribute, their efforts are often small-scale, while effective solutions would need to be scaled up on a larger scale.

Change and Resistance: Change and transformations might lead to resistance, distrust, increased conflicts, or a lack of shared vision. The process can be disorganized as groups pursue different goals. Consistent stakeholders' participation can explore different scenarios, and lead to shared visions. These theories are supported by local and global experiences. Trust is crucial in understanding a crisis, but building a trust-conducive environment is a long process with challenges, requiring stakeholders and decision-makers to manage expectations.

⁎ ⁎ ⁎ ⁎ ⁎

In an increasingly fragmented and polarized world, trust remains both a critical necessity and a fragile commodity. The barriers to trust—rooted in geopolitical rivalries, economic competition, and conflicting interests—are exacerbated by misinformation, power struggles, and institutional shortcomings. While trust is essential for fostering collaboration and addressing global challenges such as climate change, trade disputes, and security crises, its erosion deepens divisions and fuels uncertainty. Overcoming these barriers requires renewed commitment to transparency, dialogue, and shared accountability. Without such efforts, the fractures in trust risk not only undermining global cooperation but also exacerbating the very crises humanity must collectively confront.

Georg Simmel underscored that, without trust, the complexity of possible futures would paralyze action (Simmel, 1964). A generalized lack of trust in society could also lead to a "spiral of cynicism"(, Cappella et al. 1997). Are we already reaching that tipping point?

But imagine a world divided by trade conflicts where only the interests of a few would prevail—and where tensions could escalate into conflict ...

References

Braithwaite V., Levi M., 1998, *Trust and Governance*, Russell Sage Foundation series on Trust.

Brulle R. 2010, "From Environmental Campaigns to Advancing the Public Dialogue: Environmental Communication for Civil Engagement," *Environmental communication*, March 2010 4(1): 82–08.

Burch S., Shaw A. Dale A., Robinson J. "Triggering transformative change: a development path approach to climate change response in communities" in *Climate Policy*, 2014, Vol. 4, 467:487 <https://doi.org/10.1080/14693062.2014.876342>

Cappella J. N. and Jamieson K. H.,1997, *Spiral of Cynism*, New York University Oxford Press.

De Araujo Arosa Monteiro R., Ferraz de Toledo R., Roberti Jacobi R., 2021, "Dialogue Method: A Proposal to Foster Intra and Inter-community Dialogic Engagement," in *Journal of Dialogue Studies*, Special issue, Dialogue with and among the Existing, Transforming and Emerging Communities, Vol. 9, 165–183, <https://doi.org/10.55207/FWCB1722> last accessed 22/04/2024.

Deutsch M., 1958, Trust and Suspicion, Conflict Resolution Number 2 (Vol. 8).

Fine, Gary Alan, 2007, "Rumor, Trust and Civil Society: Collective Memory and Cultures of Judgment", *Diogenes* 54 (1):5–18. <https://doi.org/10.1177/0392192107073432>

Fukuyama F, 1995, *Trust: The social virtue and the creation of Prosperity* New York, Free Press.

Funtowicz S. 2020 "From risk calculations to narratives of danger," in *Climate Risk Management*, 27, 100212 (2020), <https://reader.elsevier.com/reader/sd/pii/S2212096320300024?token=B6CBEA02054999EE756B04F666D4701C3905CDF460510AD1F52DF26A57DFF692EC212672A58C4D7A49A536076F984FAB&originRegion=eu-west-1&originCreation=20210808171814>. Last accessed 08.08.2021.

Gaborit P. 2009 a, *Restaurer la confiance après un conflit civil*, L'Harmattan.

Gaborit P. 2009 b, "La confiance après un conflit ou la confiance désenchantée," in Bertho A., Gaumont-Prat H. et Serry H. *Colloque international La confiance et le conflit*, Université Paris Vincennes Saint Denis.

Gaborit P. 2021, *Learning from Resilience Strategies in Tanzania, an Outlook of International Development Challenges*, Peter Lang International. <https://www.peterlang.com/document/1152350>.

Gaborit P. 2022 (a) "Resilience and Climate Disaster Management in Cities: Transformative Change and Conflicts", *Journal of Peacebuilding & Development*, special issue, Nov. 2022, <https://doi.org/10.1177/15423166221128793>

Gaborit P. 2022 (b) "Climate Adaptation to Multi-Hazard Climate-Related Risks in Ten Indonesian Cities: Ambitions and Challenges", in *Climate Disaster Risk*, 2022 Vol. 37, 100453 <https://doi.org/10.1016/j.crm.2022.10045>

Gaborit P. 2022 (c) (Ed), *Climate Adaptation and Resilience: Challenges and Potential solutions. Anticipatory governance, Planning and Dialogue"* 2022, Peter Lang.

Giddens A. 1991, *Modernity and self-identity*, Stanford, Standford University Press.

Goodhart D. (2017) "The future to somewhere : The populist revolt and the future of politics" London, Hurst and Company.

Grafakos S., Trigg K., Landaeur M. Chelleri L., Dhakal S. 2019, "Analytical framework to evaluate the level of integration of climate adaptation and mitigation in cities," *Climatic Change*, 154: 87–106.

Greenhill K. M. and Major S. 2007, "The Perils of profiling: Civil War Spoilers and the Collapse of Intra State Peace Accords", in *International Security* Vol. 31, n°3 Winter 2006–2007, pp. 7–40.

Hallegatte, S., Vogt-Schilb, A., Rozenberg, J. *et al.* 2020, "From Poverty to Disaster and Back: a Review of the Literature." *Economics of Disasters and Climate Change* 4, 223–247. <https://doi.org/10.1007/s41885-020-00060-5> last accessed 12/04/2022.

Hansen V. D. 2019, "The case for Trump", New York Basic Books.

Hardin R. (Ed) 2002, *Trust and Trusworthiness*, New York, Russel Sage foundation editions, collection on trust, volume 4, 2002.

Hardin R. (Ed) 2004, *Distrust*, NYC, Russell Sage Foundation.

Harris L. M. Chu E., Ziervogel G. 2017, *Negotiated Resilience*, EDGES, Institute for Resources, Environment and Sustainability, University of British Colombia.

Hodge R. and Kress G. *Social Semiotics*, Cambridge, Polity, 1988.

Khodyakov D. "Trust as a process: A Three-dimensional Approach" in *Sociology*, London, 2007, vol. 41 pp. 115–132.

Lazuech G. 2002 "Toute confiance est d'une certaine manière confiance aveugle," *Pleins Feux, Variations* n°9 <https://www.librairie-sciencespo.fr/livre/9782847290004-toute-confiance-est-d-une-certaine-maniere-confiance-aveugle-gilles-lazuech/>

Lind M. 2019, *The new Class War: Saving Democracy from the Managerial Elites*, New York, Portfolio Penguin.

Livet P. 2007, "gouvernance et confiance" lecture series, Institut des nouvelles technologies de Namur, January 20, 2007.

Luhmann, N. 1979, *Trust and Power: Two Works by Niklas Luhmann. Translation of German originals Vertrauen 1968 and Macht 1975.* Chichester: John Wiley.

Marchezini V. 2020 "What is a sociologist doing here?: an unconventional People centered approach to improve warning implementation in the Sendai Framework for Disaster Risk Reduction", in *Int J Disaster Risk Sci* 11:218–229, <https://doi.org/10.1007/s13753-020-00262-1>

Marino E. and Ribot J. 2012, "Adding Insult to Injury: Climate Change and the Inequities of Climate Intervention," *Global*

Environmental Change, 22(2): 323–328. <https://doi.org/10.1016/j.gloenvcha.2012.03.001>. Last accessed 23.07.2021.

Misztal B. A. 1992, "The Notion of Trust," *Social Theory, Policy, Organisation and Society*, 5(1): 6–15. <https://doi.org/10.1080/10349952.1992.11876774>.

Möllering G. 2006, *Trust, Reason, Routine, Reflexivity*, Oxford, Elsevier.

Nyhan R. C. 2000, "Changing the paradigm Trust ad Its Role in Public Sector Organizations" in *The American Review of Public Administration* Vol. 30, March 2000 pp. 67–105.

Orléan A, 2000, "La théorie économique de la confiance et ses limites," in Laufer R. and Orillard M (dir), *La confiance en questions*, Paris, édition Harmattan, collection Logiques Sociales, 2000, pp. 59–79.

Oswald M. 2022, "The Palgrave Handbook of Populism" <https://link.springer.com/book/10.1007/978-3-030-80803-7>

Perwaiz A., Parviainen J., Somboon P., Macdonald A. 2020, "Disaster Risk Reduction in Indonesia," Status Report UN Office for Disaster Risk Reduction, Asian Disaster Preparedness Center.

Putnam R. 1993, *Making Democracy work: Civic traditions in Modern Italy*, Princeton University Press.

Putnam R. D. 1996, "The strange disappearance of civic America" in *American Prospect*,n° 24,1996 pp. 34–49.

Raven J., Stone B, Mills G, Towers J. Katzschner L, Leone M., Gaborit P. Georgescu M and Hariri M. 2018, "Climate Change and Cities: Urban Planning and Urban Design", in Rosenzweig C. et al. (eds.) *Climate Change and Cities: Second Assessment Report of the Urban Climate Change Research Network*, Cambridge University Press New York. 139–172.

Rodríguez-Pose A., "The revenge of the places that don't matter (and what to do about it)", *Cambridge Journal of Regions, Economy and Society*, Volume 11, Issue 1, March 2018, Pages 189–209, <https://doi.org/10.1093/cjres/rsx024>

Scott K. D. 1980, "The causal relationship between Trust and the assessed value of management by objectives", in *Journal of management*, vol. 6, 1980 pp. 157–175.

Seligman A., 1997, *The problem of Trust*, Princeton, Princeton University Press.

Simmel G.1964 *The sociology*, New York Free press.

Stedman S. J., 1997, "Spoiler problems in Peace Processes" in *International Security*, Vol. 22, n°2, Autumn 1997.

Stedman S. J., 2003, "Peace Processes and the challenges of violence," in Darby J. and Mac Ginty R., *Contemporary Peace Making: Conflict Violence and Peace Processes*, London and New York, Palgrave-Mac Millan editions.

Szescynski B. 1999, "Risk and Trust: The Performative Dimension," in *Environmental Values* pp. 239–252 The White Horse Press Cambridge UK.

Sztompka P., 2000, *Trust a sociological theory*, New York, Cambridge University Press.

Thuderoz C., Mangematin V. et Harrisson D. 1999, *La confiance: approches économiques et sociologiques*, Paris, édition Gaëtan Morin.

Thuderoz, C. 2003, *Négociations: essai de sociologie du lien social* Paris, PUF.

Tilly C. 2005, *Trust and Rule*, Cambridge University Press.

Van Bruggen A., Nikolic J., Kwakkel J. 2019, "Modeling with stakeholders for transformative change," *Sustainability*, 2019, 11, 825, <https://doi.org/10.3390/Su11030825>

Wheeler N. "Trust Building in international relations", Peace Prints 2011 <https://wiscomp.org/peaceprints/4-2/4.2.9.pdf>

Wilkinson W. (2019) "The Density Divide: Urbanization, Polarization, and Populist Backlash", Niskanen Center Research Paper.